Chaos in Discrete Dynamical Systems

A VISUAL INTRODUCTION IN 2 DIMENSIONS

Chaos in Discrete Dynamical Systems

A VISUAL INTRODUCTION IN 2 DIMENSIONS

RALPH H. ABRAHAM
University of California Santa Cruz

LAURA GARDINI
University of Urbino

CHRISTIAN MIRA
Institut National des
Sciences Appliquees de Toulouse

with 153 illustrations made with the assistance of Scott Hotton

Companion CD-ROM by Ronald J. Record and Ralph H. Abraham

Springer

TELOS

THE
ELECTRONIC
LIBRARY
OF
SCIENCE

Ralph H. Abraham
University of California
   Santa Cruz
P. O. Box 1378
Santa Cruz, CA 95064
USA

Laura Gardini
Instituto di Scienze
   Economiche
Università di Urbino
Urbino 61029
Italy

Christian Mira
Institut National des
   Sciences Appliquees de Toulouse
Dept. of Control Engineering
Toulouse 31077
France

Publisher: Allan M. Wylde
Editor: Paul Wellin
Publishing Associate: Keisha Sherbecoe
Product Manager: Walter Borden
Production Supervisor: Natalie Johnson
Manufacturing Supervisor: Jeffrey Taub
Production: Jan V. Benes, Black Hole Publishing Service
Copy Editor: Paul Green
Cover Designer: Irene Imfeld

Library of Congress Cataloging-in-Publication Data
Abraham, Ralph.
    Chaos in discrete dynamical systems : a visual introduction in
  2 dimensions / Ralph H. Abraham, Laura Gardini, Christian Mira.
       p.   cm.
    Includes bibliographical references and index.
    ISBN 0-387-94300-5 (hardcover : alk. paper)
    1. Differentiable dynamical systems.   2. Chaotic behavior in
  systems. I. Gardini, L. (Laura) II. Mira, C. III. Title.
  QA614.8.A268 1997
  003'.857—dc21   96-37581

Photocomposed copy prepared using the authors' electronic files by
Black Hole Publishing Service, Berkeley, CA.

Printed and bound by Hamilton Printing Co., Rensselaer, NY.

Printed in the United States of America.

9 8 7 6 5 4 3 2 1

ISBN 0-387-94300-5   Springer-Verlag New York Berlin Heidelberg   SPIN 10470118

TELOS, The Electronic Library of Science, is an imprint of Springer-Verlag New York with publishing facilities in Santa Clara, California. Its publishing program encompasses the natural and physical sciences, computer science, mathematics, economics, and engineering. All TELOS publications have a computational orientation to them, as TELOS' primary publishing strategy is to wed the traditional print medium with the emerging new electronic media in order to provide the reader with a truly interactive multimedia information environment. To achieve this, every TELOS publication delivered on paper has an associated electronic component. This can take the form of book/diskette combinations, book/CD-ROM packages, books delivered via networks, electronic journals, newsletters, plus a multitude of other exciting possibilities. Since TELOS is not committed to any one technology, any delivery medium can be considered. We also do not foresee the imminent demise of the paper book, or journal, as we know them. Instead we believe paper and electronic media can coexist side-by-side, since both offer valuable means by which to convey information to consumers.

The range of TELOS publications extends from research level reference works to textbook materials for the higher education audience, practical handbooks for working professionals, and broadly accessible science, computer science, and high technology general interest publications. Many TELOS publications are interdisciplinary in nature, and most are targeted for the individual buyer, which dictates that TELOS publications be affordably priced.

Of the numerous definitions of the Greek word "telos," the one most representative of our publishing philosophy is "to turn," or "turning point." We perceive the establishment of the TELOS publishing program to be a significant step forward towards attaining a new plateau of high quality information packaging and dissemination in the interactive learning environment of the future. TELOS welcomes you to join us in the exploration and development of this exciting frontier as a reader and user, an author, editor, consultant, strategic partner, or in whatever other capacity one might imagine

TELOS, The Electronic Library of Science
Springer-Verlag Publishers
3600 Pruneridge Avenue, Suite 200
Santa Clara, CA 95051

THE ELECTRONIC LIBRARY OF SCIENCE

TELOS Diskettes

Unless otherwise designated, computer diskettes packaged with TELOS publications are 3.5" high-density DOS-formatted diskettes. They may be read by any IBM-compatible computer running DOS or Windows. They may also be read by computers running NEXTSTEP, by most UNIX machines, and by Macintosh computers using a file exchange utility.

In those cases where the diskettes require the availability of specific software programs in order to run them, or to take full advantage of their capabilities, then the specific requirements regarding these software packages will be indicated.

TELOS CD-ROM Discs

For buyers of TELOS publications containing CD-ROM discs, or in those cases where the product is a stand-alone CD-ROM, it is always indicated on which specific platform, or platforms, the disc is designed to run. For example, Macintosh only; Windows only; cross-platform, and so forth.

TELOSpub.com (Online)

Interact with TELOS online via the Internet by setting your World-Wide-Web browser to the URL: *http://www.telospub.com.*

The TELOS Web site features new product information and updates, an online catalog and ordering, samples from our publications, information about TELOS, data-files related to and enhancements of our products, and a broad selection of other unique features. Presented in hypertext format with rich graphics, it's your best way to discover what's new at TELOS.

TELOS also maintains these additional Internet resources:
gopher://gopher.telospub.com
ftp://ftp.telospub.com

For up-to-date information regarding TELOS online services, send the one-line e-mail message:
send info to: *info@TELOSpub.com.*

Dedicated to
Igor Gumowski

FOREWORD TO THE PROJECT

You are looking at the outcome of a three-year project, a unique experiment in electronic publishing. For lack of a better word, we call this a *package.* It has three intertwined components: a *book,* a *CD-ROM,* and a *website.* It is perhaps the first such multimedia package devoted to an advanced branch of mathematics.

The book is the primary component, and it is extensively illustrated with monochrome computer graphics. The CD-ROM is devoted mainly to twelve computer graphic animations in color, which animate and expand the graphics in the book. The user interface to the CD-ROM is made in the style, and with the technology, of the World Wide Web. Therefore, it integrates seamlessly with the website devoted to the book and CD-ROM, which is maintained at the Visual Math Institute. This website also connects outward with the resources of the World Wide Web.

The motivation for this package is the conviction that this style of electronic publication is the ideal medium for mathematical communication, and especially for the branch of mathematics known as dynamical systems theory, including our subject: noninvertible discrete chaos theory in two dimensions. The essence of this communicative style is the *dynapic* technique, in which a drawing is developed stroke-by-stroke, along with a carefully coordinated spoken commentary. This is the traditional method used by most mathematicians, when speaking among themselves: *Visual Math!*

We will now introduce the three components separately.

Ralph H. Abraham
Santa Cruz, California
October 1996

PREFACE TO THE BOOK

This book is a visual introduction to chaos and bifurcations in noninvertible discrete dynamical systems in two dimensions, using the method of critical curves.

Historical Background

Dynamical systems theory is a classical branch of mathematics which began with Newton circa 1665. It provides mathematical models for systems which evolve in time according to a rule, originally expressed in analytical form as a system of ordinary differential equations. These models are called *continuous dynamical systems.* They are also called *flows,* as the points of the system evolve by flowing along continuous curves.

In the 1880s, Poincaré studied continuous dynamical systems in connection with a prize competition on the stability of the solar system. He found it convenient to replace the continuous flow of time with a discrete analogue, in which time increases in regular, saltatory jumps. These systems are now called *discrete dynamical systems.* So, for over a century, dynamical systems have come in two flavors: continuous and discrete. Discrete dynamical systems are usually expressed as the iteration of a map (also called an endomorphism) of a space into itself. In these systems, points of the system jump along dotted lines with a regular rhythm.

In the context of a discrete dynamical system, in which a given map is iterated, that map might be *invertible* (because of being one-to-one and onto) or *noninvertible* (failing one or the other or both of these conditions). So, discrete dynamical systems come in two types, invertible and noninvertible. The invertible maps were introduced by Poincaré, and have been extensively studied ever since. The studies of noninvertible maps have been more sparse until recently, when they became one of the most active areas on the research frontier because of their extraordinary usefulness in applications.

Chaos theory is a popular pseudonym for dynamical systems theory. This new name became popular about 20 years ago, when its applicability to chaotic systems in nature became widely known through the advent of computer graphics. As there are two flavors of dynamical systems, continuous and discrete, there are also two chaos theories. The first to develop, in the work of Poincaré about a century ago, was the theory of chaotic behavior in continuous systems. He also studied chaotic behavior in discrete dynamical systems generated by an invertible map.

Discrete chaos theory for noninvertible maps began some years after Poincaré. Its development has been accelerated particularly since the computer revolution, and today it is a young and active field of study. The earliest development of the theory came in the context of *one-dimensional maps,* that is, the iteration of a real function of a single real variable. One of the key tools in the one-dimensional theory was the calculus of critical points, such as local maxima and minima. The *two-dimensional context* is the current research frontier, and it is the subject of this book.

For two-dimensional noninvertible maps, the *critical curve* is a natural extension of the classical notion of critical point for one-dimensional noninvertible maps. The first introduction of the critical curve, as a mathematical tool for two-dimensional noninvertible maps, appeared in papers by Gumowski & Mira in the 1960s (see the bibliographies at the end of the book for references.)

The importance of our subject

Chaos theory is crucially important in all the sciences (physical, biological, and social) because of its unique capability for modeling those natural systems which behave chaotically. It is for this reason that there is a chaos revolution now ongoing in the sciences. For those systems which present continuous, evolving, data (such as the solar system), continuous chaos theory provides models. And for those which present discrete data (such as economics), discrete chaos theory provides models. One advantage of discrete dynamical models is the ease and speed of simulating the models with digital computers, as compared with continuous dynamical models. Discrete models are sometimes advantageous, even in the context of natural systems presenting continuous data.

Uniqueness of this publication

The book component of this book/CD-ROM/Website package is not a conventional text book, and yet its purpose is pedagogic. It intends to provide any interested person possessing a minimal background in mathematics, but with a basic understanding of the language of set-theory, an introduction to this new field. It is unique in providing both an elementary and a visual approach to the subject. While chaos theory is mathematically sophisticated, by focusing on examples and visual representations — there are about one hundred computer graphics in the book — and minimizing the symbols and jargon of formal mathematics which they are relegated to a set of appendices, the text provides the reader with an easy entry into this important and powerful theory. The primary focus of the package is the concept of *bifurcation* for a *chaotic attractor*. These are introduced in four exemplary bifurcation sequences, each defined by a family of very simple noninvertible maps of the plane into itself. Each family, the subject of an entire chapter in the book, exhibits many bifurcations.

And as dynamics involves motion, computer graphic animations provide a particularly appropriate medium for communicating dynamical concepts. The CD-ROM contains twelve animations which bring life to the basic ideas of the theory, literally animating the still images of the book. For each of the four map families there is one long, fast movie which is a fast forward through the entire chapter, as well as two "zooms" which expand a brief piece of the action into a slow motion movie. The movies can be understood only by reading along in the book while viewing the movie. The motion controls of the movie players (in both the Windows and the Macintosh environments) allow easy stop, play, fast-forward, reverse, and slow-motion, by dragging a slider. This makes the CD-ROM ideal for studying in conjunction with the book.

Intended audience

While many devotees of pure mathematics may enjoy this package for the novelty of its fresh ideas and the mathematical challenge of a new subject, the intended audience for this book is the large and heterogeneous group of science students and working scientists who must, due to the nature of their work, deal with the modeling and simulation of data from complex dynamical systems of nature which are intrinsically discrete. This means, for example, applied scientists, engineers, economists, ecologists, and students of these fields.

How to use the book

The book is divided into three nearly independent parts. The first part provides the simplest introduction to the basic concepts of discrete chaos theory, with many drawings and examples. The second part is a detailed analysis of computer experiments with four families of discrete chaotic systems, with emphasis on the method of critical curves, and the phenomena of bifurcation. The third part is a set of appendices which provide more official definitions for readers who have a stronger background in abstract mathematics. Extensive historical material by Professor Mira is also found here, some made available in English for the first time. It is proposed that the second part be regarded as a "guided tour" through a very difficult terrain, and each example studied repeatedly, with reference as necessary (using the index) to the first and third parts, and to the CD-ROM.

ABOUT THE BOOK AUTHORS

Ralph H. Abraham is Professor of Mathematics at the University of California at Santa Cruz, founder of the graduate program there on computational dynamics, and is an author of

- *Foundations of Mechanics,*
- *Manifolds, Tensor Analysis, and Applications,* and
- *Dynamics, the Geometry of Behavior.*

Laura Gardini is Professor of Mathematics at the Universities of Urbino and Brescia in Italy, and is an author of

- *Chaotic Dynamics in Two-Dimensional Noninvertible Maps.*

Christian Mira is Professor of Control Engineering at the Institut National des Sciences Appliquees de Toulouse, in France, a founder of a laboratory of computational dynamics there, and is an author of

- *Dynamique chaotique: transformations ponctuelles, transition ordre-desordre,*
- *Recurrences and discrete dynamic systems,*
- *Chaotic Dynamics,* and
- *Chaotic Dynamics in Two-Dimensional Noninvertible Maps.*

As the creator of the method of critical curves, Christian Mira brings to this book long and extensive experience in the field. Laura Gardini extended the method of critical curves and applied it extensively, recently obtaining many new results. Ralph Abraham, known for his pioneering work, and his extensive book writing and illustrating, on continuous dynamical systems since 1960, met Mira and Gardini at a conference in June, 1991, and soon after became their co-author on this work.

PREFACE TO THE CD-ROM

The CD-ROM supplied in the back of the book is intended as an enhancement to the book. Its main function is to animate the graphics in Chapters 4 through 7 with twelve movies. It also contains some useful software. This companion CD-ROM may be regarded as a "canned" piece of the World Wide Web. It has an index which may be accessed by any WWW browser, like Netscape Navigator, or Microsoft Internet Explorer. The CD also connects seamlessly with the Web if your computer has Internet access.

The movies

The movies for Chapters 4, 5, 6, and 7 are computer graphic animations, created by extensive computations with ENDO, an X-Windows software package developed for research on discrete dynamical systems in two dimensions by Ronald Joe Record. These movies provide the best opportunity to understand the role of critical curves in the bifurcations presented in these chapters.

Each of the four chapters — 4, 5, 6, and 7 — present an exemplary bifurcation sequence. This means that we are given a one-parameter family of maps, and we carefully observe a chaotic attractor as the parameter is varied. Certain special events called bifurcations occur, perhaps very frequently, as the parameter is changed. In each of these chapters, we have singled out just a few of these special events for special attention, which we call "stages".

For example, in Chapter 4, there are twelve stages. In the book, monochrome computer graphics are included for each of these stages, along with extensive commentary which tries to explain the very complicated images.

In the movies, the stages are embedded in a very large number of in-between images, which are then flashed on the screen like a flip book. Thus, the still-frame black-and-white stage images of the book are embedded in an apparently continuous, uniform sequence of color-coded images in the movies. The color code is a one-dimensional spectral scale from blue to red, and is shown at the right side of the screen in all of the movies. In the square frames of the movies, the color blue indicates a low relative density of trajectory points in a given small square of the plane, while red indicates a high density.

Additional CD-ROM content

Besides the twelve movies, each in two formats, the CD-ROM also contains additional material: MAPLE and ENDO.

The 96 computer graphics in Chapters 4 through 7 of the book (with four exceptions) have been computed in the mathematical programming language MAPLE by Scott Hotton. For the 92 images that have been made in this way, the complete programs (they are plain text files) may be read directly from the CD-ROM. Reading one of these files, with the help of a MAPLE programming manual if needed, answers all possible questions about the figures in the book: the size of the domain, the number of points, and other considerations. In addition, the programs are very easily modified and run in the MAPLE environment to do further research in chaos theory.

The ENDO program, written by Ron Record, was used by him to make all of the frames of the movies on the CD-ROM. It is an easy-to-use research environment which you might use to do frontier research in two-dimensional discrete chaos theory, if you have access to an X-Windows environment. We have included the complete program on the CD-ROM, in an archived and compressed UNIX file. Instructions for its installation are found in the file "index.htm" on the CD-ROM.

Finally, the CD-ROM contains (in file "index.htm") a few pointers to relevant websites, for those who have an Internet connection.

How to navigate the CD-ROM

There are two methods for accessing the CD-ROM.

Method #1. The first method, which we strongly recommend, makes use of a World Wide Web browser. The one we have used is Netscape Navigator, which is freely available on the Internet. All other browsers should work, but we have not tested them. In this method,

    a. Insert the CD-ROM in the CD-ROM drive.
    b. Start the browser.
    c. Click the File item on the browser menu bar.
    d. Choose the "Open File" option.
    e. Browse to the file "index.htm" on the CD-ROM.
    f. Open it.

All contents of the CD-ROM are then displayed for your choice. This is particularly convenient for the MAPLE script files. Also, if you happen to be connected to the World Wide Web, you may click on some links to external servers.

Note: Clicking on a movie in the web browser results in a one-minute wait, while the movie file is copied from the CD-ROM to your hard disk. Accept the wait as a necessary inconvenience, because the movies play better from the hard disk (unless your equipment is in perfect running order). After the wait, you will see the first frame of the movie in the web browser window. You may then start and stop the movie by clicking anywhere in its frame.

Method #2. This is the fall-back method, and does not require any software other than the Windows FileManager, Macintosh Desktop, or UNIX shell.

a. View the contents of the CD-ROM.
b. Double click on the item of choice.

Because this CD-ROM is a hybrid CD, the file structure maintains the look of whatever operating system it is on.

Hardware and software requirements

The twelve movies are each provided in two formats on the CD-ROM: AVI and QuickTime. Both are 320x240x8 video with 22kHz by 16 bit sound. On a Macintosh, you must use the QuickTime versions. Under Windows you would choose the AVI version, unless you have QuickTime for Windows on your system, in which case you have a choice. QuickTime for Windows is available from Apple over the Internet, and our CD-ROM has a link to Apple to help you obtain a copy. In any case, you may play the movies through the web browser, as described earlier in the preferred Method #1. On the other hand, with the fall-back Method #2, the QuickTime movies may be played with the Movie Player included in the Macintosh operating system, while the AVI movies may be played with the MediaPlayer which is part of the Windows operating system.

These movie players have a simple control panel with run and pause buttons. In addition, you may drag the slider to advance or reverse the movie at slower or faster than normal speeds. You may use either format on a UNIX platform, with appropriate software, such as the freeware "xanim" for X-Windows. On Windows or Macintosh machines, you may also use a World Wide Web Browser to play the movies, as we have explained previously.

The movies assume that your computer is capable of playing QuickTime (MOV) or Video for Windows (AVI) movies at 2X speed, that is, at 300KB per second. If the movies jerk or stick, that probably means that your computer needs a tune-up.

Bugs

Every hardware/software platform plays CD-ROMs differently, and we cannot anticipate all of the potential problems. We have tested our CD-ROM on several machines of each sort; Windows, Macintosh, and UNIX. All functions have been robust and correct except the movie service function.

On older versions of Windows and Macintosh operating systems, the movie players seem to occasionally stick. As a workaround, try moving the slider back and forth to loosen things up. Some older systems display a warning message upon first inserting the CD-ROM in its drive, but <RETURN> seems to work.

Here are some tricks to improve Macintosh movie performance. **Virtual Memory:** Typically, this is set on, and to about 1MB more than the actual RAM. For example, with actual RAM 16MB, set virtual RAM to 17MB. **Cache Memory:** This may be reduced to improve movie playing. **MoviePlayer Application Memory:** Increase the amount of memory devoted to MoviePlayer if you know how.

ABOUT THE CD-ROM AUTHORS

Ralph H. Abraham created the computational dynamics program at the University of California at Santa Cruz.

Ronald J. Record is a Ph.D. graduate of the computational dynamics program at the University of California at Santa Cruz, and now works as a software engineer in Santa Cruz.

PREFACE TO THE WEBSITE

All of the material currently available is found in the book, or on the CD-ROM. However, upon publication of this package, additional graphics, questions and answers will be posted on the web site devoted to the project and administered by the Visual Math Institute. We will maintain a Chaos FAQ (Frequently Asked Questions) and bug reports on the site, and other features which may prove useful to the international chaos community.

The URL for the web site is: http://www.vismath.org/chaos/jpx

ACKNOWLEDGMENTS

Thanks to Raymond Adomaitis, Gian Italo Bischi, Robert Devaney, and Daniel Lathrop, for their very helpful comments on the text, and Karen Acker for her FrameMaker artistry. Our publisher, TELOS, has been extraordinarily supportive and patient with our process: a thousand thanks to Allan Wylde, Paul Wellin, and Jan Benes. And we are deeply indebted to Scott Hotton for his extraordinary illustrations in the book and corresponding MAPLE programs on the CD-ROM.

A video by John Dorband of NASA Goddard Space Flight Center originally sparked our collaboration presented in Chapter 7 here, and we are grateful to him for sharing his work with us.

Figures 7-24, 7-25, 7-28, and 7-29, very difficult to compute, are the work of our colleague Daniele Fournier-Prunaret, and we are grateful to her for contributing them to this book. Her beautiful drawings have inspired us.

Finally, we are very grateful to Peter Broadwell for the loan of a Silicon Graphics Indigo computer, which ran continuously for several *months* cranking out the frames for the movies on our CD-ROM. And without our mathematical copy editor, Paul Green, this book would be a mine field for the novice reader. We are deeply in his debt.

CONTENTS OF THE BOOK

CONTENTS OF THE CD-ROM

Preface to the CD-ROM (Readme.txt)
Index (index.html)
Copyright notice (copyright.htm)

Movies, AVI and MOV formats:

Movie 4-1:   Ch. 4, full sequence of 7 stages.
Movie 4-1a: Ch. 4, 20X zoom of stage 6.
Movie 4-1b: Ch. 4, 200X zoom of stage 7.
Movie 5-1:   Ch. 5, full sequence of 11 stages.
Movie 5-2:   Ch. 5, 10X zoom of stages 2 through 5.
Movie 5-3:   Ch. 5, 10X zoom of stages 8, 9, 10.
Movie 6-1:   Ch. 6, full sequence of 12 stages.
Movie 6-2:   Ch. 6, 10X zoom of stages 5, 6, 7.
Movie 6-3:   Ch. 6, 20X zoom of stages 8 through 12.
Movie 7-1:   Ch. 7, Full sequence of 12 stages.
Movie 7-2:   Ch. 7, 20X zoom of stages 1 through 7.
Movie 7-3:   Ch. 7, 20X zoom of stages 8, 9, 10.

Additional material:

MAPLE scripts for figures in the book
The ENDO program</A>

Web site pointers:

Visual Math Institute, the website for this package.
TELOS, our publisher
Netscape, to download Navigator
Microsoft, to download Internet Explorer
Apple, to download QuickTime
Brooks-Cole, to obtain MAPLE
Laboratory of Prof. Mira, for recent research

PART I

BASIC CONCEPTS

In the first three chapters we present an overview and the basic concepts, beginning with critical points in the one-dimensional case and proceeding to the corresponding idea of critical curves in two dimensions.

CHAPTER I

INTRODUCTION

1.1 BACKGROUND

Our goal is to present the fundamentals of two-dimensional (2D) iteration theory through examples, with extensive graphics (for which the 2D context is ideal) and few mathematical symbols.[1] We illustrate all the basic ideas with hand drawings and monochrome computer graphics in the book, and again with movies (full-motion video animations in color) on the companion CD-ROM.

We do not assume a knowledge of higher mathematics. But we do acknowledge that our subject is a branch of pure mathematics, and a deeper understanding requires some topology and geometry. A hint of this is presented in the appendices, where a more rigorous approach is introduced.

1.2 HISTORY

The study of chaos in 1D iterations is a classical subject, going back to Poincaré over a century ago, as described in detail in Appendix 5. The 2D case (two real variables or one complex variable) also goes back almost a century but the stream of literature to which this book belongs really begins with the computer revolution and the pioneers of scientific computation — Von Neumann, Ulam, and so on — in the 1950s. Our subject remains an experimental domain, and computer graphic experiments provide our main orientation. Our fundamental tool for describing the behavior of 2D iterations, the *critical curve*, was introduced by Gumowski and Mira in 1965.[2]

---

1. This approach was developed in (Abraham, 1992).
2. See (Gumowski and Mira, 1965) and (Mira, 1965) in the Bibliography.

1.3 PLAN OF THE BOOK

In Part 1, we introduce the basic concepts and vocabulary of iteration theory, first in 1D, then in 2D. We try to introduce only as much theory as is required to understand Part 2, on exemplary bifurcation sequences. In Part 2, we will use the vocabulary and ideas of Part 1 to explain step-by-step the events in the exemplary bifurcation sequences.

We use the critical curves to understand the structure of attractors, basins, basin boundaries, and their bifurcations. Then we increase the bifurcation parameter, and explain the changes in the configuration of attractors and basins due to bifurcations of various types, again using the critical curves.

These structures and changes are illustrated with still images created by our software for the iteration of a fixed endomorphism, based on the method of critical curves. These graphics are strung together as movies, which may be viewed from the accompanying CD-ROM, to give a more dynamic idea of the sequence of bifurcation events in each of the exemplary families.

1.4 CONTEXT

Dynamics is a vast subject, and our subject is a relatively new frontier within it. So, for those who already have an idea of the territory of dynamical systems, we would like now to locate our subject within this larger territory.

Dynamical systems theory has three flavors:

- *flows* are continuous families of invertible maps generated by a system of autonomous first-order ordinary differential equations, and parameterized continuously by time, that is, by real numbers;

- *cascades* are discrete families of invertible maps generated by the iteration of a given invertible map, and parameterized discretely by the integers (zero, positive, and negative);

- *semi-cascades* are discrete families of maps generated by iteration of a given map, generally noninvertible, and parameterized discretely by the natural numbers (zero and the positive integers).

Both cascades and semi-cascades are also known as *discrete dynamical systems*, or *iterations*. In this book we are primarily interested in semi-cascades generated by a noninvertible map, (NIM). For simplicity, we will simply call these iterations in future; but keep in mind that all of this book belongs to the NIM flavor.

In general, the *state space*, the space in which a dynamical systems is defined, may be an arbitrary space of any dimension: 1, 2, 3, and so on. This suggests a tableau of types of dynamical systems, as shown in Figure 1-1. In this tableau, there is a relationship between cells on the same diagonal (marked with an A): In each row, the marked cell is the cell of lowest dimension in which *chaos* occurs. Hence, the tableau is called the *stairway to chaos*. Here chaos means any dynamic behavior more complicated than periodic behavior.

In this book, we discuss only the iteration of noninvertible maps, and the only state spaces we consider are the one-dimensional Euclidean line and the two-dimensional Euclidean plane. In fact, the latter is our primary subject. The 1D case has been extensively treated in recent literature (see M1) and shares the stairway to chaos with 2D cascades and 3D flows, the contexts for the early history of chaos theory. (See Appendix 5.) The second diagonal, marked with B here, may be regarded as the current frontier of chaos theory.

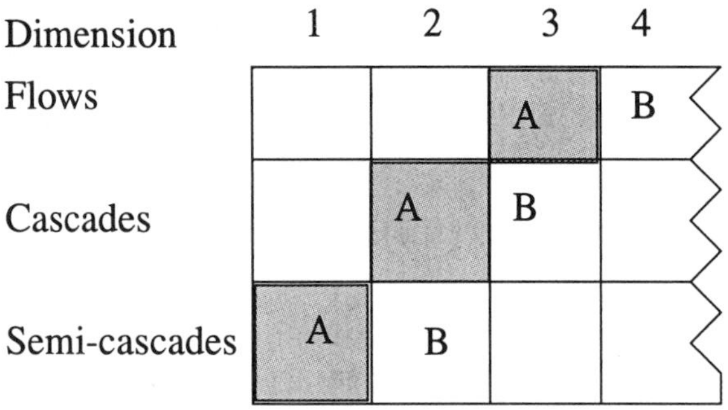

FIGURE 1-1.

The stairway to chaos.

Dimension	1	2	3	4
Flows			A	B
Cascades		A	B	
Semi-cascades	A	B		

1.5 BASIC CONCEPTS OF ITERATION THEORY

This section introduces the basic terminology. These concepts will be explained in detail in 1D and 2D in the next two chapters.

Iterated map: An iterated map is the generator of a discrete dynamical system; generally a noninvertible, continuous map.

Multiplicity: Our maps are usually *noninvertible*, that is, many-to-one, so a given point may have several preimages. The range set may be decomposed into with zones of constant multiplicity (bounded by *critical points* or *critical curves*) in which all points have the same number (called the *multiplicity*) of preimages. These multiple preimages determine a tree of *partial inverses* for the map. Multiplicities (explained further in the next chapter) play a fundamental role in our theory, analogous to the degree of a polynomial function.

Critical sets: These sets are boundaries of zones of constant multiplicity; thus, they separate zones of different multiplicity. They consist of points with *coincident inverses*.

Zones: The zones of constant multiplicity play a very fundamental role in our view of NIM theory, analogous to the role of degree of a polynomial in algebra.

Partial inverses: By restricting our attention to a zone of constant multiplicity in the range of a map, say multiplicity k, we may define k inverses to the map. These partial inverses play the role, in the NIM context, of the inverse of an invertible map.

Trajectory: A trajectory embodies the basic data of a dynamical system. It consists of the list of locations of the images of a particular point, called the *initial point*, under the iterations of the map generating the dynamical system. It is an ordered sequence (as opposed to a set) of points.

Attractors, basins, boundaries: These are the chief characteristic features of an iteration, from the qualitative point of view. Attractors are limit sets of trajectories of initial points filling an open set, which is the basin of the attractor. The boundaries of these basins are particularly important in applications of the theory.

Portrait: The state space of a dynamical system may be decomposed into a set of open sets (basins), in each of which is a single attractor. The boundaries of these basins are particularly important in applications of dynamical systems theory.

Bifurcations: These are fundamental changes in the qualitative behavior of a dynamical system, occurring as a control parameter is varied. At certain critical values of the parameter, the qualitative behavior of the trajectories of the system suddenly changes in a significant way. These sudden changes, called bifurcations, usually occur in sequences, called *bifurcation sequences*.

1.6 THE FAMILIES OF MAPS

The first family of maps we use to illustrate the basic ideas of discrete dynamics is from a paper of Kawakami and Kobayashi, studied also in a paper of Mira and coworkers:

$$u = ax + y$$

$$v = b + x^2 \qquad \text{EQ 1}$$

Usually, we fix the value of *a*, and vary *b* to exhibit a *bifurcation sequence*. We make use of three special cases in Part 2:

- Case 1: a = 0.7 (Chapter 4, Absorbing Areas)

- Case 2: a = 1.0 (Chapter 5, Holes)

- Case 3: a = -1.5 (Chapter 6, Fractal Boundaries).

These parameter ranges are shown in Figure 1-2.

In Chapter 7, Chaotic Contact Bifurcations, we use the double logistic map studied by Gardini and coworkers,

$$u = (1-c)x + 4cy(1-y)$$

$$v = (1-c)y + 4cx(1-x)$$

EQ 2

with *c* in [0, 1].

FIGURE 1-2.

Parameter space of the first family.

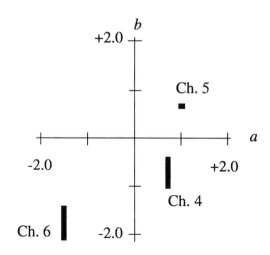

1.7 CRITICAL POINTS AND CURVES

The theory of critical curves for maps of the plane provides powerful tools for locating the chief characteristic features of a discrete dynamical system in two dimensions: the location of its chaotic attractors, its basin boundaries, and the mechanisms of its bifurcations. We next introduce the basic concepts of this theory first in 1D, then in 2D.

CHAPTER 2

BASIC CONCEPTS IN 1D

In the preceding chapter we introduced a brief list of basic concepts of discrete dynamics. Here, we expand on these concepts in the one-dimensional context, in which, uniquely, we have the advantage of a simple graphical representation. The official, abstract definitions of all these concepts may be found in the Appendices.

2.1 MAPS

By a *map* we mean a continuous function from a space, called the *domain*, to itself.[1] In the one-dimensional context, the domain might be an interval (with or without endpoints) of the real number line, or even the entire line.

If f is a map on a real interval, I, we indicate this in symbols $f:I \rightarrow I$. For example, if the map is defined by the rule $f(x) = x^2$, and I is the closed interval [-2, 2], we may visualize the map graphically, as shown in Figure 2-1. The action of the map is to move points from the horizontal axis to the vertical axis in two strokes:

- vertically from the horizontal axis to the graph,

- horizontally from the graph to the vertical axis,

as shown in Figure 2-2.

The *image* (or *range*) of the map is the set of all points obtained as $f(x)$ while x takes on all values in the domain, I, and is written in symbols as $f[I]$. For a point y in I, a *preimage of rank 1* is a point x

---

1. The word *continuous* belongs to the branch of math known as *point-set topology*. This wonderful subject is not known as well as it ought to be, but nevertheless, we must use it constantly.

in I that is mapped to y; that is, x is a preimage of rank 1 of y if $y = f(x)$. A *preimage of rank 2* of y is a preimage of rank 1 of a preimage of rank 1, and so on. Every point y in I has a set of preimages of every rank, which may be empty. Determining all preimages of a point creates a genealogical tree, called the *arborescent sequence of preimages*.

The map f is *one-to-one*, if, for every point y in I, the set of preimages of y has either no points or just one point. For example, the map f defined by the same rule, $f(x) = x^2$, but with the smaller domain $[0, 1]$, is one-to-one. A one-to-one map has a unique inverse map, $f^{-1}: f[I] \to I$ which undoes what f does. This can be visualized on the graph of f as a motion in two strokes:

- horizontally from the vertical axis to the graph of f,

- vertically from the graph to the horizontal axis,

as shown in Figure 2-3.

Note that if we try to invert the map of Figure 2-2 by this two-stroke process, we discover all of the preimages of a given point y in I (represented as the vertical axis) in one step. This is shown in Figure 2-4, where we find two preimages.

In this text we will be concerned exclusively with the lowest step of the staircase to chaos, the 1D case. We will be interested especially with maps which are not one-to-one. These are called *many-to-one*, or *noninvertible*, maps. For such maps, points generally have more than one preimage of rank 1, and the number of preimages of a given rank determines a zone of multiplicity, discussed below.

2.2   MULTIPLICITIES

Given any map of an interval I, we may choose a point y in I on the vertical axis, locate all preimages by the graphical method, and count them up. Thus, we may decompose the vertical axis into sets of points all sharing the same number of preimages. We denote these zones by:

FIGURE 2-1.

Graph of *f* on
I = [-2, 2].

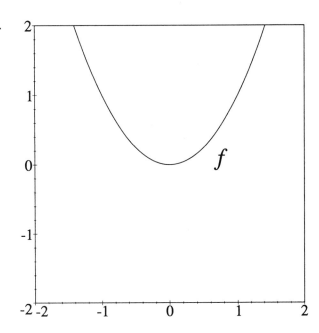

FIGURE 2-2.

Two strokes from a
point *x* to its image *y*.

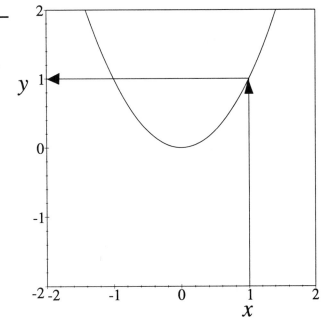

- Z_0 (all points having no preimages)

- Z_1 (all points having exactly one preimage)

- Z_2 (all points having two distinct preimages)

and so on.

These are called the *layer sets* of the map; those sets that are nonempty and open (that is, have no boundary points) are called *multiplicity zones*. The zones Z_0 and Z_2 are shown in Figure 2-5. They exhaust the whole interval $I = [\_2, 2]$ except for the point 0, which separates the two zones. We also say this map is *of type $Z_0 - Z_2$*, meaning there are two zones, one of multiplicity zero, the other of multiplicity two. The suffix indicates the multiplicity. The interval Z_2, may be considered a folded image, that is, two halves of the domain I are folded onto this image. For each half of the domain, our map does have an inverse. These are called *partial inverses*.

The point 0 is the only point of Z_1 in this example; that is, it has multiplicity one (its unique preimage is 0). It is called a *critical point* because it lies on the boundary of two zones. In fact, we can describe this map as a *nonlinear folding*. That is, the map folds the horizontal axis at the critical point, then stretches them in a nonlinear fashion onto the range interval. This is why the critical point is sometimes called a *fold point*.

Generally, we will be interested in relatively simple maps, such as polynomials, in which only finite multiplicities, with generic (that is, typical) fold points, are encountered. We call these *finitely folded maps*. For example, a typical cubic map has multiplicities 1 and 3, and we say it is of type $Z_1 - Z_3 - Z_1$. These basic concepts of iteration theory should be approached through a careful study of simple (for example, polynomial) examples.

2.3 TRAJECTORIES AND ORBITS

A map generates a discrete dynamical system by iteration. That is, the map is applied again and again, and points move along a dotted path called a *trajectory*. For example, choosing an *initial point* x_0, let x_1 denote its image under the map f, likewise x_2 the image of x_1, and so on. The infinite sequence $(x_0, x_1, x_2,...)$ is the trajectory

---

CHAOS IN DISCRETE DYNAMICAL SYSTEMS

FIGURE 2-3.

Two strokes from a
point y to its preimage
of rank 1.

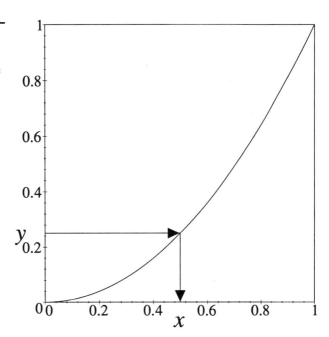

FIGURE 2-4.

Finding all preimages
of rank 1 of a point y.

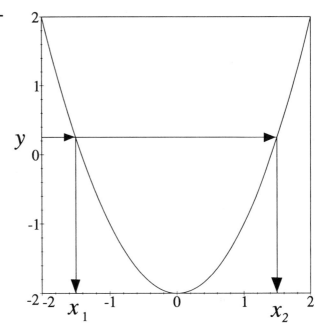

of x_0. This sequence may jump around a finite set of points. The minimum set of points which holds a trajectory its called its *orbit*. When finite, an orbit is called *cyclic*, or *periodic*, and the number of its points is its *order*. A *fixed point*, defined by $f(x) = x$, is a special kind of periodic point, the orbit of which is a single point. It has order 1. If an orbit contains only two points, it is called a 2-*cycle*, and so on.

We now describe a graphical method for plotting trajectories, called the *Koenigs-Lemeray method*. Note that in our graphs, both axes represent the same set, since the domain and range of our map consist of the same interval.

Given x_0, envisioned on the horizontal axis, we may find x_1 on the vertical axis by the two-stroke method described in 2.1. Next, we must repeat this process, starting from the point x_1 on the horizontal axis. Our immediate problem, then, is to transfer the distance x_1 from the vertical axis to the corresponding distance on the horizontal axis.

One way to carry out this transfer is shown in Figure 2-6. Here we use a compass to measure the vertical distance, x_1, and rotate it to the horizontal distance, x_1. Another method is to use a protractor to construct a line descending at slope -1, or 45 degrees, as shown in Figure 2-7.

Yet another method — and this is the one we prefer — is shown in Figure 2-8. We draw a line from the lower left corner, ascending at slope 1. This line is called the *diagonal* (in symbols, Δ). Now, using only a square, we draw a horizontal line from x_1 on the vertical axis until it meets Δ, then draw a vertical line until it meets the horizontal axis. This determines the horizontal distance, x_1, as shown in Figure 2-8.

The entire construction from horizontal x_0 to vertical x_1 to horizontal x_1 may now be summarized as follows:

- vertical from horizontal axis to graph,

- horizontal from graph to vertical axis,

- horizontal from vertical axis to diagonal,

- vertical from diagonal to horizontal axis.

FIGURE 2-5.

The multiplicity zones
for a quadratic function
on [-2, 2].

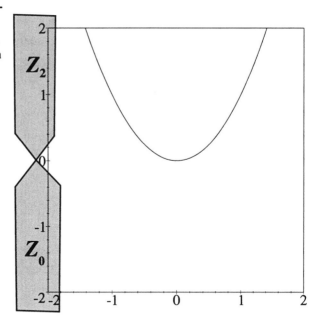

FIGURE 2-6.

The compass
construction.

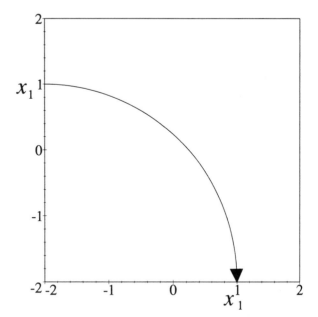

FIGURE 2-7.

The descending line method.

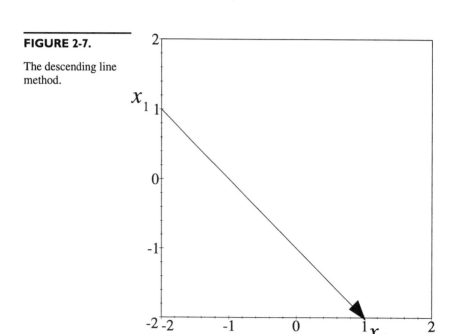

FIGURE 2-8.

The square two-stroke method.

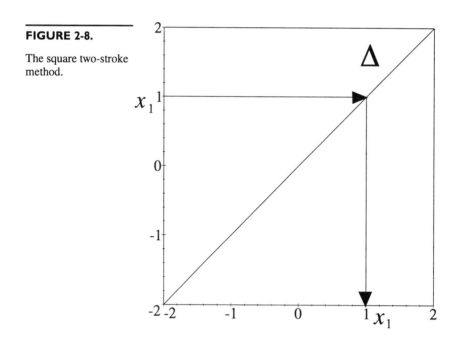

This construction may be abbreviated somewhat since the third stroke retraces (undoes) part of the second, as shown in the four strokes of Figure 2-9. The abbreviated construction (Figure 2-10) is:

- vertical from horizontal axis to graph,
- horizontal from graph to diagonal,
- vertical from diagonal to horizontal axis.

This is the three-stroke graphical method for plotting one step of a trajectory within the horizontal axis. When we proceed to plot the point x_2 by this method, however, we find a further opportunity for abbreviation. The last stroke above, which locates x_1 on the horizontal axis, *i.e.*,

- vertical from diagonal to horizontal axis,

is followed by the first stroke of the second step,

- vertical from horizontal axis to graph,

which may be combined into a single stroke,

- vertical from diagonal to graph.

Thus the iterated sequence, beginning with x_0 on the horizontal axis, is:

- vertical from horizontal axis to graph,
- horizontal from graph to diagonal,
- vertical from diagonal to graph,

and continue. After the first step, we may pretend that we are jumping about on the diagonal, which after all is just another copy of the domain interval I. Each step has two strokes,

- vertical to the graph; horizontal to the diagonal

(which we may remember by the mnemonic, *vertigo-horrid*), as shown in Figure 2-11. That is the graphical method of Koenigs-Lemeray, also known as the *staircase method*, or *cobweb construction*. Using it, we may quickly follow trajectories for several jumps on the diagonal. See Figure 2-12.

BASIC CONCEPTS IN ID

FIGURE 2-9.

The four strokes from a
point on the horizontal
axis to its image on the
horizontal axis.

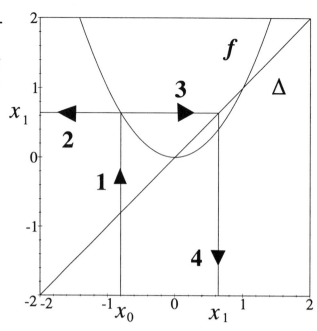

FIGURE 2-10.

The abbreviated three-
stroke method.

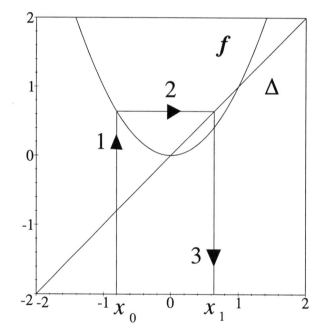

FIGURE 2-11.

One step, consisting of two strokes on the diagonal.

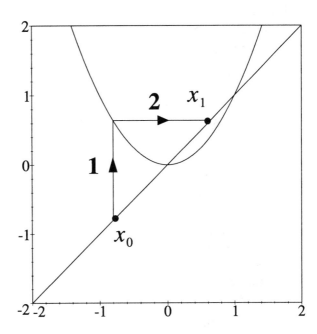

FIGURE 2-12.

The staircase method of Koenigs and Lemeray.

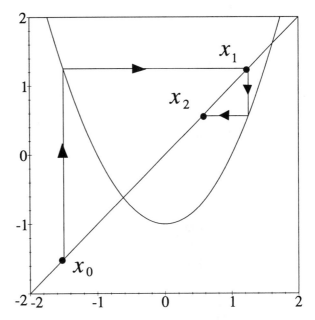

Using this method, one may graphically verify this useful fact: if a point returns to its starting point after two iterations of the map, the starting point is either a fixed point or a 2-periodic point of the map.

2.4   ATTRACTORS, BASINS, AND BOUNDARIES

Try out the staircase method using the *Myrberg map*,[1] $f(x) = x^2 - c$, with the entire real line as the domain, and various values for the control parameter, *c*. You may quickly find that some trajectories converge to a *fixed point*, while others run off to positive infinity (upper right) on the diagonal. The fixed points are seen immediately as the crossing points of the graph and the diagonal, and are defined by the property: $f(x) = x$.

An interval is called *trapping* if it is mapped into itself, and *invariant* if it is mapped exactly onto itself. If a bounded interval is trapping, then all of its trajectories are trapped inside, and must converge to a closed, invariant, and bounded limit set. These limit sets are the *attractors* of the map. Attractors may be classified in three categories:

- a *point attractor* is a single point,

- a *cyclic attractor* is a finite set of points, and

- a *chaotic attractor* is any other type of attractor.[2]

The *basin* of an attractor is the set of all points tending to that attractor. The domain is decomposed into the basins of different attractors, including the *basin of infinity*, which consists of all points whose trajectories run away from any bounded set.

The *boundaries* of the basins, also called *frontiers* or *separatrices*, are of primary importance in dynamical systems theory. A detailed study of a map results in a *portrait*, in which the domain is decomposed into basins, one attractor shown in each.

---

1. Myrberg was one of the first to study the bifurcation sequence of this map. See the Bibliography for references to his work.

2. This reflects the fact that different definitions of chaos abound in the literature.

2.5 BIFURCATIONS

As in the Myrberg map, $f(x) = x^2 - c$, we frequently encounter maps which depend on a parameter. As the parameter is changed, the portrait of the attractive set of the map may change gradually and insignificantly; however, as certain special values of the parameter are crossed, there may be a sudden and significant change in the portrait of the map. These special values are called *bifurcation points*, and the sudden changes in the portrait are called *bifurcations*. At the present time, dynamical systems theory does not have a satisfactory and rigorous definition of bifurcation, but the subject is now evolving through the study of examples. In fact, the goal of this book is to describe some of these examples, in a two-dimensional context.

In the current one-dimensional context, we again have the benefit of an excellent visualization device, the *response diagram*. In the case of a single control parameter, this is a two-dimensional graphic in which the vertical axis represents the domain of the map and the horizontal axis represents the control parameter. Above each point on the horizontal axis, the portrait of the corresponding map is indicated, with its attractors, basins, and basin boundaries. For a one-parameter family of maps of a two-dimensional domain, the response diagram is three-dimensional, as we will soon see.

2.6 EXEMPLARY BIFURCATION

The simplest bifurcations are the *fold* and the *flip*. These may involve changes to any kind of attractor. To introduce the basic concepts of bifurcation theory, however, we will describe the fold bifurcation in the simplest case, which involves point attractors.

The fold bifurcation is a *catastrophic bifurcation*. This means that, as the control parameter varies, an attractor appears or disappears suddenly. In this event, as shown in Figure 2-13 with the control parameter moving to the right on the horizontal axis, a fixed point appears, and immediately separates into a pair of distinct fixed points. One is an attractor, the other, a repellor. The repellor is shown below the attractor. Points between the two fixed points are attracted to the upper fixed point, and repelled by the lower fixed point. These tendencies are indicated by the arrows in Figures 2-15 to 2-17.

To understand the mechanism of this bifurcation, we now turn to a specific example, the Myrberg family of maps, $f(x) = x^2 - c$. The graph of a map of this family is an upward-opening parabola, with the vertex on the vertical axis at distance c below the horizontal axis. As c increases, the parabola moves downward. Three cases of this graph are shown in Figs. 2-14, 2-15, and 2-16.

In the first case, Figure 2-14, with $c = -0.5$, the parabola does not meet the diagonal because for this value of c, there are no fixed points. All trajectories tend upward without bound, to infinity.

In the next case, Figure 2-15, with $c = -0.25$, the parabola meets the diagonal in a single point, which is the fixed point $x = 0.5$, corresponding to this value of c, the bifurcation value. Trajectories approach from below, but depart from above.

In the last case, Figure 2-16, with $c = 0$, the parabola cuts the diagonal in two points, the fixed points $x = 0$ and $x = 1$, which are, respectively, an attractor and a repellor.

The flip is a *subtle bifurcation*. This means that, in contrast to catastrophic and explosive bifurcations, its effect is too subtle to observe at the moment of bifurcation when the control parameter passes its critical value, but becomes apparent later, as the parameter continues to increase. In the flip, a point attractor loses its attractiveness. From it is emitted a cyclic attractor of period 2.

A response diagram of this event is shown in Figure 2-17. To the left of the bifurcation value of the control parameter (horizontal axis) there is a single fixed point, and it is an attractor, *FP+*. The attraction in the vertical state space is shown by the heavy arrows. To the right, there is still only one fixed point, but it is a repellor, *FP-*. But there is also a 2-cycle, which is attractive, *2P+*. Looking only at the attractors in the picture, we see that the attractive point has been replaced by an attractive 2-cycle, as the control parameter moves to the right. At first, the two points of this 2-cycle are very close together, then they gradually separate.

We now move on to two dimensions.

FIGURE 2-13.

The response diagram of a fold bifurcation. The state space, where the dynamics occur, is vertical. The control parameter, c, is horizontal. The parabolicoid curves locate the fixed points of the maps. The bifurcation occurs, in the Myrberg example, when $c = -0.25$.

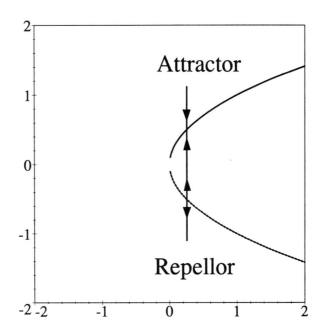

FIGURE 2-14.

A typical member of the Myrberg family, before the fold bifurcation. The graph of the map is entirely above the diagonal, so there are no fixed points.

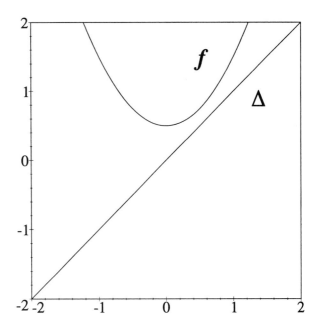

FIGURE 2-15.

At the fold bifurcation. The graph of the map has made contact with the diagonal at a single fixed point.

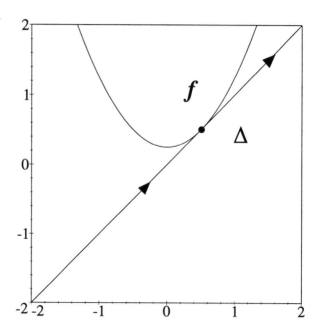

FIGURE 2-16.

After the fold bifurcation. The graph now meets the diagonal in two points, both fixed.

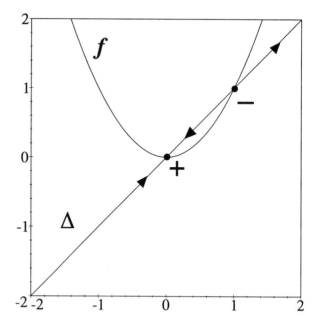

FIGURE 2-17.

The response diagram
of the flip bifurcation.

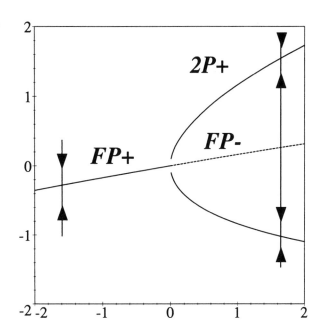

CHAPTER 3

BASIC CONCEPTS IN 2D

The basic concepts named in the Introduction, and described in the preceding chapter in a 1D context, apply with little modification in the 2D context which is our main concern in this book. We no longer have the convenience of a visible graph of the map, however, because the graph of a 2D map is a 2D surface in a 4D space. Therefore, we must be satisfied with a frontal view of the 2D domain of the map, in which we try to visualize as much as possible.

3.1 MAPS

As before, by *map* we mean a continuous function from the *domain* to itself, $f:D \to D$. From now on, the domain will be a two-dimensional subset, usually an open subspace, of the plane. For example, D might be an open rectangle (that is, not containing its boundary) or the whole plane. The *images* and *preimages* of a point, the *one-to-one* property, and *noninvertibility* are defined as in 2.1. We now consider noninvertible maps, in 2D.

Note: The complex number maps familiar from the fractal theories of Fatou, Julia, Mandelbrot, and others may be regarded as real 2D maps. Thus, they fit in the context of this book.

3.2 MULTIPLICITIES AND CRITICAL CURVES

In the context of a given map, we define the *layer set*, Z_n, as the set of points having exactly n preimages of rank 1, where n is a natural number: 0, 1, 2, ..., and so on. Those layer sets that are nonempty open sets are the *multiplicity zones*. Points on the boundaries of the zones, generally, are *critical points* of the map. In general, these sets will not exhaust the domain, as there may be points that have infinitely many preimages of rank 1, and thus do not belong to Z_n for any n. Even polynomial maps may have this problem, but generic (that is, almost all) polynomial maps have the following nice properties:

- there are only a finite number of layer sets;

- they exhaust the domain;

- the zones of multiplicity fill almost all of the domain;

- all layer sets that are not multiplicity zones consist of critical points, arranged in a set of piecewise smooth curves.

A map having these nice properties is called a *finitely folded map*, and a curve consisting of critical points is called a *critical curve of rank 1*, and is denoted by L.[1] The image of a critical curve of rank 1 is a critical curve of rank 2, denoted L_1, and so on.

Note: For a given map T, $L = T[L_{-1}]$, where L_{-1} is the *critical curve of rank 0*, and may be thought of as the set of "coincident preimages" of points of L.

3.3   AN EXAMPLE

For example, let D be the entire plane. The polynomial map $f: D \to D$ defined by $f(x, y) = (u, v)$, where,

$$u = ax + y$$

$$v = b + x^2$$

EQ 3

is finitely folded. We will set $a = -0.7$, and $b = 1.0$.

Figure 3-1 shows part of the domain of the map, the (x, y) space D, containing the critical curve of rank 0, L_{-1}, which coincides with the y axis. It divides the domain into two regions, denoted R_1 and R_2. Figure 3-2 shows part of the image of the map f, in the (u, v)

---

1. In the original literature, L is usually denoted by LC, for the French term, *ligne critique*. More rigorous definitions may be found in Appendix 3.

FIGURE 3-1.

The domain of (x, y) divided by the critical curve, $x = 0$.

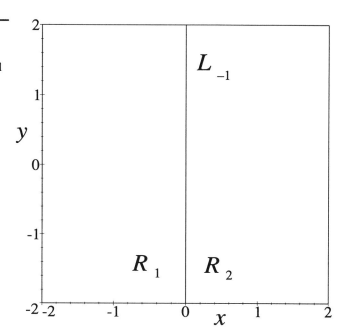

FIGURE 3-2.

The range of (u, v) divided by the critical curve, $v = b = 1.0$. The image of the map is above L.

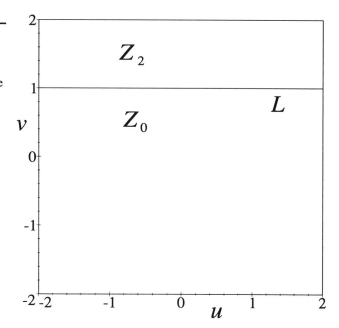

space, with the zones of multiplicities zero and two, Z_0 and Z_2, separated by the critical curve, L. Figure 3-3 shows the two spaces, superimposed.

There is a *folding* of the (x, y) space, on the critical curve of rank 0, L_{-1}, followed by a nonlinear deformation, a rotation, and a movement to the right, into the space of (u, v). The entire (x, y) space ends up on the zone Z_2, with the critical curve L_{-1} moving onto the critical curve, L. We may visualize the two regions of the (x, y) space, R_1 and R_2 folded onto one another, then distorted and pressed down onto Z_2. Actually, the two regions are mapped onto one.

The motion, visualized in this way, may be reversed. This provides a method to visualize the action of the inverse mapping as well. A small area in the (u, v) space on the right, if contained entirely within Z_2, will unfold into two small regions in the (x, y) space on the left, one in the region R_1, the other in R_2.

As we wish to iterate the map, and to visualize the trajectories, attractors, and basins, of our discrete dynamical system, it will be useful (although initially confusing) to superimpose the (u, v) space on top of the (x, y) space. Then, as a weak substitute for the graphical method of Koenigs-Lemaray in the 1D case, we apply the

FIGURE 3-3.

The image of the map, superimposed on the domain.

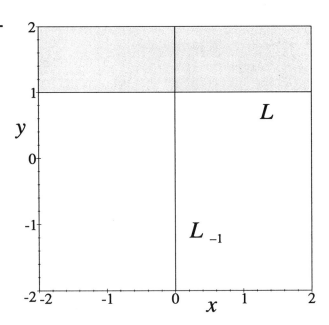

motion from Figure 3-1 to Figure 3-2 again and again. The domain, D, is mapped into itself repeatedly. The curve L_{-1} moves onto the curve L, which in turn moves onto the curve L_1, and so on. This superimposition is shown in Figure 3-3. This portrait is the basis of the *method of critical curves*, which is the main method of this book.

3.4   TRAJECTORIES AND ORBITS

In the 2D context, *trajectories* and *orbits* are defined exactly as in 2.3 but here they must be plotted in the two-dimensional domain. This is particularly appropriate for computer-generated plots. And in this computational method, the *attractors* and their *basins* may be discovered by experiment. We usually find the critical curve of rank 0, L_{-2} manually by the standard method of vector calculus (involving the vanishing of the Jacobian determinant, see Appendix 3), then enter its symbolic description into the computer program, which can then plot the higher-order iterates. The method of critical curves is based on experiments such as this.

As in the 1D case, there are special kinds of orbits which are important qualitative features of the dynamics of an iterated map. First among these are the *fixed points*, which are unmoved by the map. The different types of fixed points are defined by the motions of nearby points. The classification is based on the differential calculus and linear algebra of two-dimensional spaces, but we will give here only the results. Excepting certain unusual cases, there are five kinds of so-called *generic* fixed points in this classification. The five types are illustrated in Figures 3-4 to 3-8.

Another special type of orbit of great importance in the qualitative theory is the *periodic orbit*, or *cyclic orbit*, or *cycle*, which consists of a finite set of points. The map permutes the points of the orbit cyclically: If there are n points in the orbit, each of the points returns to its original position after exactly n iterations of the map. The number n is called the *period* of the orbit, which is also called an *n-cycle*. A point of an n-cycle is said to have *prime period n*, and is also a fixed point of the map iterated n times. Periodic points are classified according to their type as a fixed point of the iterated map.

FIGURE 3-4.

Attractive focus. All
nearby points are
attracted and spiral
toward the fixed point.

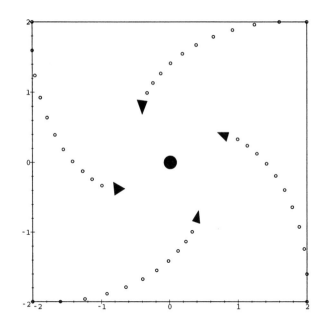

FIGURE 3-5.

Attractive node. All
nearby points are
attracted, and tend to
approach along a curve
through the fixed point.

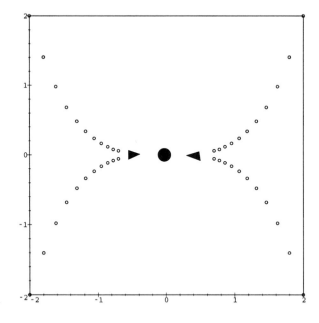

FIGURE 3-6.

Saddle. A repellor, most nearby points are attracted, and then repelled along a curve through the fixed point.

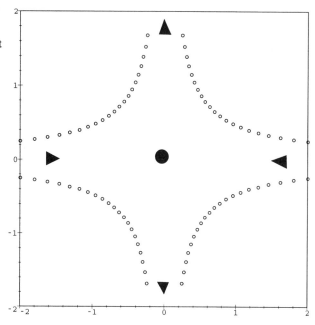

FIGURE 3-7.

Repelling node. All nearby points are repelled, and tend to depart (at least briefly) along a curve through the fixed point. The opposite of an attractive node.

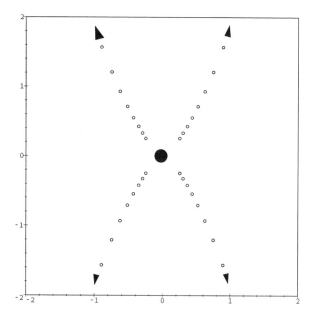

FIGURE 3-8.

Repelling focus. All nearby points depart, spiralling away from the fixed point. The opposite of an attractive node.

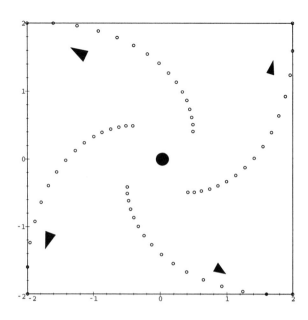

3.5   ATTRACTORS

As in the 1D case, there are three types of *attractors*:

- *static attractors*, also called *attractive fixed points*;

- *periodic attractors*, also called *cyclic attractors*; and

- *chaotic attractors*.

The static attractors are fixed points which are attractive, that is, the trajectories of all nearby points are attracted to them. Of the five types of fixed points illustrated in Figures 3-4 to 3-8, two are attractive, and three repelling. Periodic attractors are periodic orbits (orbits of trajectories that cycle around a finite point set) which are attractive.

Chaotic attractors are more complicated sets which are attractive. For the mathematically inclined, technical definitions are given in Appendix 2. For others, these concepts will gain meaning through examples later in the book.

3.6   BIFURCATIONS

The informal definition of *bifurcation*, in 2D, is the same as in 1D. Again, there are subtle and catastrophic bifurcations, and other distinctions such as local versus global bifurcations. These are best understood in the examples dissected in detail in the following chapters.

In the 2D context, a one-parameter family of maps may be displayed in a *response diagram*, in which the bifurcations may be seen and analyzed. This is a 3D plot in which the domain of the maps, a 2D set, is arrayed vertically, and moved along a horizontal axis representing the control parameter. In each of these vertical planes, the portrait of attractors, basins, and boundaries must be visualized. In practice, this is a challenging task of computer graphics, and we usually seek a simpler display. The technique we adopt for this book, which is well accommodated by computer graphic animation technology and CD-ROM media, is the animated movie. Thus, we translate the control parameter into the time dimension and view the domain of the map head on, watching the attractor-basin portrait adjust itself to a time-changing control parameter.

The method of characteristic curves becomes a strategy for the analysis of these bifurcation movies. So, on to the exemplary bifurcation sequences.

PART 2

EXEMPLARY BIFURCATION SEQUENCES

A bifurcation sequence is a sequential occurence of intersting bifurcation events in a family of maps, seen as the parameter is varied.

In the next four chapters we describe four particularly exemplary sequences. Each is taken from the recent literature and illustrates a basic type of bifurcation behavior. Each chapter is devoted to one of these types, and may begin with an introduction to the basic concepts in 1D and 2D. The four types of bifurcation are:

- absorbing areas,
- holes,
- fractal boundaries, and
- contact bifurcations.

CHAPTER 4

ABSORBING AREAS

We begin with a brief introduction to the concept of absorption in one and two dimensions, and then study an exemplary bifurcation sequence.

4.1 ABSORPTION CONCEPTS, 1D

We introduced in Chapter 2 the notions of critical points, which bound zones of multiplicity, and trapping intervals, which are mapped into themselves. These notions come together in the concept of an *absorbing interval*. This is an interval in the domain of the map which is trapping, is bounded by critical points, and is *super-attracting*, which means that every point sufficiently close to the critical endpoints will jump into the absorbing interval after a finite number of applications of the map.

In the context of an iterated map of an interval, the interesting dynamics take place within absorbing intervals. The critical points of a one-dimensional map determine absorbing intervals, and are useful in characterizing some bifurcations, especially those called global bifurcations. Examples of global bifurcations occur in Chapter 7.

4.2   ABSORPTION CONCEPTS, 2D

In two-dimensional iterations, we have a notion of *absorbing area*, generalizing the absorbing intervals of the one-dimensional case. The critical curves of a map of the plane play a role analogous to that of the critical points of a one-dimensional map: they are useful in determining absorbing areas. We will now illustrate the role of critical curves in determining these important areas in which the interesting dynamics occur.

In this chapter we will study the first family of quadratic maps defined in the Introduction. We will now use the map of this family determined by $b = -0.8$ to illustrate the use of critical curves to determine an absorbing area.

Our map has two fixed points, P and Q. The *basic critical curve*, L, is the locus of points with "coincident preimages," that is, the set of points having nearby points with different numbers of preimages or rank 1 (see Appendix 3, especially A3.2 – A3.3, for definitions). The *fundamental critical curve L_{-1}* is defined as the preimage of L. Thus, L is the image of L_{-1} under the map. Similarly, the *derived critical curve L_1* is the image of L, and so on. All of these are called *critical curves*, as described in Chapter 2.

Note for those who have studied vector calculus: In the context of a generic smooth map, the fundamental critical curve L_{-1} will be a subset of the set of *critical points* in the Jacobian sense, points at which the Jacobian derivative of the map (a linear transformation) is degenerate (not a linear isomorphism), while the basic critical curve L is a subset of the set of *critical values* in the Jacobian sense. *Inflection points* are Jacobian critical points which do not belong to L_{-1}.

For this particular map, L_{-1} is the vertical axis, $x = 0$, and L is a horizontal line, $y = b = -0.8$. It will be convenient to choose a bounded interval in L_{-1}, S_{-1}, with endpoints a_{-1}, which is the origin $(0, 0)$, and a_0, which is the image of a_{-1} under the map, the point $(0, -0.8)$.

Note: The critical curve denoted by L in the text is denoted by L_0 in the figures.

Figure 4-1 shows all these features and more. Note the points a_{-1} and a_0, and the segments of L_{-1}, L, L_1,..., L_4. As the map moves L_{-1} to L, and the point a_{-1} to a_0, L is moved to L_1, and the point a_0 to a point a_1 in L_1. The transversal[1] (and here, indeed, orthogonal) crossing of L_{-1} and L at a_0 is transformed into a tangent contact of L and L_1 at a_1. Then this point is mapped to a

---

1. An intersection of two curves is said to be *transversal* if they cross cleanly through each other in a single point, and are not tangent to each other.

tangency of L_1 and L_2 at a_2, and so on. Note that the curve L_2 crosses L_{-1} at the point p_0, so L_3 is tangent to L at the point p_1, the image of p_0, and so on. Similarly, the curve L_3 crosses L_{-1} at the point b_0, so L_4 is tangent to L at b_1, the image of b_0, and so on.

It is worthwhile to pause here and carefully study Figure 4-1. A point of transversal crossing of any curve, C, through L_{-1} is mapped into a point of tangency of the image of that curve, $f(C)$, with L. This is because of the *folding* which occurs as L_{-1} is mapped onto L. Also, a point of tangency of a curve C to the curve L_{-1} is mapped into a point of tangency of the image curve $f(C)$ and L.

The action of the map may be visualized as a nonlinear folding of the plane at the fundamental critical curve, L_{-1}, followed by a nonlinear rotation moving L_{-1} to L around the point a_0, and a nonlinear translation horizontally along L, so that a_{-1} ends up at a_0. The critical curves shown in Figure 4-1 are a kind of skeleton of the map, as we shall see.

FIGURE 4-1.

An absorbing area
bounded by critical
curves.

a=0.7   b=-0.8

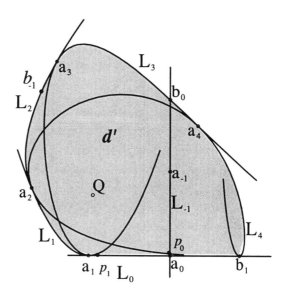

Our next goal is to use these critical curves to discover an absorbing area of the map. By definition (see Appendix A3.5), an absorbing area is a region of the plane such that:

- it is mapped into (or onto) itself;

- its boundary is made up of segments of critical curves, or of limit points of an infinite sequence of critical curves; and

- it has a neighborhood every point of which eventually moves into the absorbing area.

As an example, note in Figure 4-1 that the arcs $b_1 a_1$ of L, $a_1 a_2$ of L_1, $a_2 a_3$ of L_2, $a_3 a_4$ of L_3, and $a_4 b_1$ of L_4 bound a region d', shown shaded in Figure 4-1. This shaded region happens to be an absorbing area! First of all, it is invariant. For example, the segment $a_4 b_1$ of L_4 is on the boundary of the region, d', but the image of this arc, $a_5 b_2$, is internal to d'. This is clearly shown in Figure 4-2, in which the image of $a_4 b_1$ is indicated in L_5.

Also, d' is absorbing: A point external to d' is mapped into d' in a finite number of iterations, as shown in Figure 4-3, and this is the case for all points sufficiently near to d'.

FIGURE 4-2.

An annular absorbing area bounded by critical curves.

a=0.7   b=-0.8

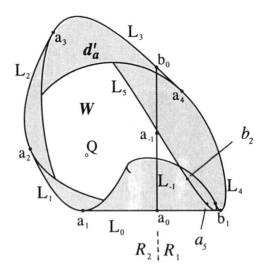

Figure 4-2. also shows a shaded area, d'_a, bounded by the critical curve segments introduced in Figure 4-1. It surrounds a hole, W, in which the fixed point Q is located. This is a smaller absorbing area, and is called an *annular absorbing area* for obvious reasons. Generally, absorbing areas may be topologically more complex, with many holes, and with many separate pieces.

Absorbing areas must contain attractors. One, d, is shown as a cloud of dots in Figure 4-3. It is the attracting set of d', and of the smaller absorbing area, d'_a, as well.

Usually there are smaller and smaller absorbing areas around any attractor of the map. All these, by definition, are bounded by critical arcs. And as they get smaller and smaller, but always enclose the same attractor, it is to be expected that the attractor itself is bounded by critical arcs, or by limit points of infinite sequences of critical arcs. Figure 4-4 shows 25 iterates of two intervals of L_{-1}, the two pieces of $d \cap L_{-1}$. These iterates bound the attractor shown in Figure 4-3 quite closely.

Another basic concept of dynamics is the basin of an attractor. In the method of critical curves, we usually determine an absorbing area by the method illustrated above, and thus the attractor within it, and then determine the basin of attraction of the absorbing area. In Figure 4-5 the basin of attraction, $D(d')$, is shown as the white region. Its boundary is made up of the repelling (nodal) fixed point, P, the saddle 2-cycle $\{Q_1, Q_2\}$, and its insets.[1] The points of the gray region go off to infinity. That is, their trajectories are unbounded. We call this set the basin of infinity, $D(\infty)$. The white area $D(d')$ (excluding the fixed point Q and its rank 1 preimage Q_{-1} in Z_0) is the basin of the attractor d.

4.3   EXEMPLARY BIFURCATION SEQUENCE

In this first exemplary bifurcation sequence, we use the first family with $a = 0.7$, and decrease b from -0.4 to -1.0, in seven stages.[2] For these values of b, our map always has two fixed points, P and Q, given by:

1. By *inset* of P we mean the set of points attracted to P, also called the *stable set* of P.

2. We follow the paper BB.

FIGURE 4-3.

The attractor within
the annular absorbing
area.

a=0.7   b=-0.8

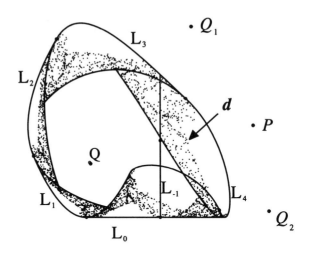

FIGURE 4-4.

Critical curves con-
verging to the
boundary of the
attractor.

a=0.7   b=-0.8

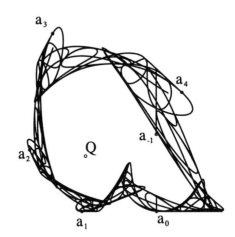

CHAOS IN DISCRETE DYNAMICAL SYSTEMS

FIGURE 4-5.

The attractor in its basin.

a=0.7   b=-0.8

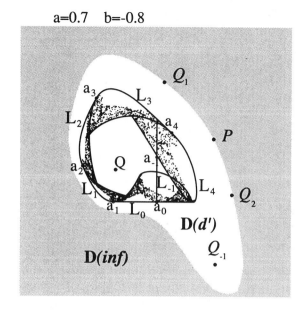

$$P: x = \frac{(1-a+\sqrt{\delta})}{2}, \ y = (1-a)x$$

$$Q: x = \frac{(1-a-\sqrt{\delta})}{2}, \ y = (1-a)x$$

where $\delta = (1-a)^2 - 4b$. Also, there is a 2-cycle, $\{Q_1, Q_2\}$.

Stage I: b = – 0.4

Here, the point Q, at about $(-0.5, -0.15)$, is an attractive fixed point. As b decreases, a Neimark-Hopf bifurcation occurs. The fixed point Q becomes a repellor and the curve Γ appears as an attractive invariant cycle, Γ (a closed curve mapped onto itself) that gradually increases in size as b continues to decrease.

Stage 2: b = – 0.5

In Figure 4-6, we see the point repellor Q within the attractive cycle Γ, and surrounding that, our first absorbing area, d', shown shaded in Figure 4-7. This area is mapped into itself, is bounded by arcs of critical curves, and is attractive (see Appendix 3.5 for the precise definition). In this case, the absorbing area is bounded by the arcs of the critical curves, L, L_1, L_2, L_3, and L_4. These bounding arcs are generated by successive iterations of the map, as we now describe.

Notice in Figure 4-7, which shows the critical arcs in more detail, that L_{-1} and L are straight lines, crossing orthogonally in one point. Let a_0 denote this point, $(0, – 0.5)$. Since it belongs to L, it has a unique preimage, a_{-1}, in L_{-1}. In this case a_{-1} is the origin $(0, 0)$.

Keeping the interval $a_{-1} a_0$ in mind, we now draw the successive images of the halfline of L_{-1} issuing from a_{-1} and containing a_0, that is, issuing downwards. The fourth image, lying within L_3, crosses the interval $a_{-1} a_0$, as shown in Figure 4-7. At this event our constructive procedure ends, we have found an absorbing area, shown shaded in this figure. Further successive images of the segment converge on Γ, as shown in the blowup, Figure 4-8. This procedure may be called *Procedure 1*. A related procedure is illustrated in the next stage.

As b continues to decrease, Γ expands further and eventually crosses L_{-1}.

Stage 3: b = – 0.6

In this case Γ crosses L_{-1} in the two points p_0 and q_0. as shown in Figure 4-9. The wavy shape of Γ is a consequence of this crossing, and may be understood as follows.

Apply the map to the configuration shown in Figure 4-9. The points p_0 and q_0 are mapped into the points p_1 and q_1, also on Γ. The straight line segment of L_{-1} between p_0 and q_0 is mapped into the straight line segment of L between p_1 and q_1. Because the map folds the plane two-to-one while moving L_{-1} to L, the curved segment of Γ between p_0 and q_0 is carried into the curved segment of Γ between p_1 and q_1, which is above the line L. The

FIGURE 4-6. a=0.7 b=-0.5

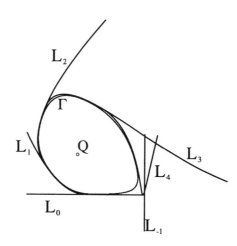

FIGURE 4-7. a=0.7 b=-0.5

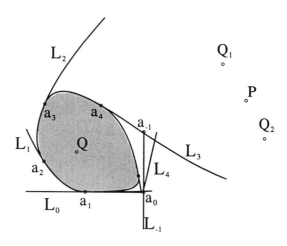

FIGURE 4-8.

a=0.7   b=-0.5

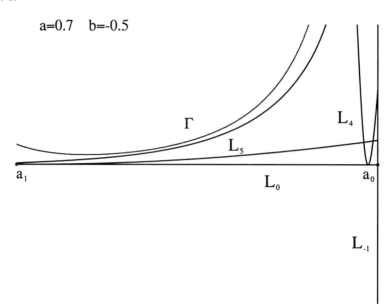

FIGURE 4-9.

a=0.7   b=-0.6

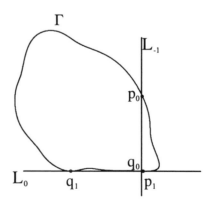

transversal crossings of Γ through the line L_{-1} are mapped into tangencies of Γ with L at the points p_1 and q_1. These tangencies are shown clearly in the enlargement, Figure 4-10. This is a universal property of curves crossing L_{-1}: All crossings are mapped into tangencies, or contacts, because the map folds the plane at L_{-1} and maps the two sides of L_{-1} onto just one side of L. Hence, Γ obtains a wave from the image of its segment which has crossed L_{-1}.

Another effect of these tangencies is that Γ is now tangent to the boundary of the absorbing area d' identified in Stage 2 above, as shown in Figure 4-11. This absorbing area may be found by the following method, called *Procedure 2*.

Consider the straight line segment S_{-1} from a_{-1} to a_0 in L_{-1} as above, and construct its successive images $a_m a_{m+1}$ by repeated applications of the map, until the first crossing with L_{-1}, in the point b_0. See that the first image of S_{-1}, S_0, is a straight line segment from a_0 to a_1 in L. The second image, S_1, is a wave from a_1 to a_2 in L_1, likewise S_2 in L_2, S_3 in L_3, and S_4 in L_4. But S_4 crosses L_{-1}, and thus b_0 is found. Let A_{-1} denote the straight line segment from b_0 to a_0 in L_{-1}. Then A_{-1} contains the segment S_{-1} constructed just above, and its image A_0 is a straight line segment from b_1 to a_1 in L, containing S_0. Now the curve segments A_0, S_1, S_2, S_3, B enclose the absorbing area d', where B is the curve segment from $a_4 b_1$ within S_4 and L_4, as shown in Figure 4-12.

At this stage, we may see yet another absorbing area, which is annular in shape. That is, it has a hole. This is shown bounded by shaded curves in the enlargement of Figure 4-13. Its boundary is constructed of successive images of the straight line segment A_{-1} from b_0 to a_0 in L_{-1}. The attractive invariant curve, Γ, is tangent to the external boundary of this annular absorbing area, as well as to its interior boundary.

As b decreases further, many bifurcations occur in the dynamics within these absorbing areas. Probably they have not all been discovered, but we show just a few events in the remaining figures of this chapter.

FIGURE 4-10.

a=0.7 b=-0.6

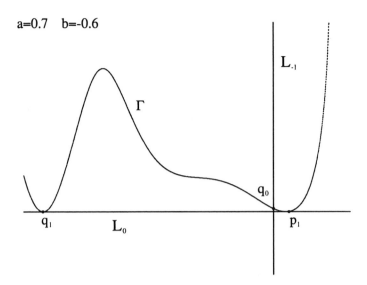

FIGURE 4-11.

a=0.7 b=-0.6

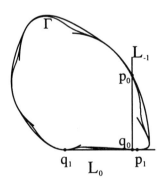

FIGURE 4-12.

a=0.7   b=-0.6

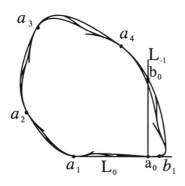

FIGURE 4-13.

a=0.7   b=-0.6

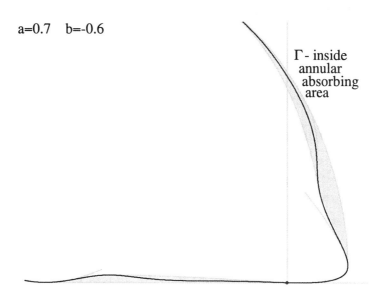

Γ- inside
annular
absorbing
area

Stage 4: b = – 0.72

At this stage there is an attractive periodic cycle of period 11. The points of this cycle are labelled in iteration sequence in Figure 4-14. This 11-cycle persists as b decreases further, through very many more bifurcations. The movie on the CD-ROM reveals an astonishing number of these, and many more have been observed, even in an interval of b values as narrow as 0.001.

Stage 5: b= –0.78

At this stage we find two attractors coexisting within the annular absorbing area, a 28-cycle and an 11-piece chaotic attractor. These are shown in Figure 4-15. Near $b = 0.798$, there is an explosion to a chaotic attractor filling an annular absorbing area.

FIGURE 4-14.

The attractive 11-cycle. Note the permutation sequence indicated by the numbers.

a=0.7   b=-0.72

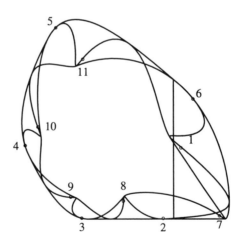

Stage 6: b = − 0.7989995

Figure 4-16 shows the chaotic attractor, bounded by critical curves. Passing below $b = − 0.8$, there are a number of additional bifurcations which have been studied on the research frontier. Some of them will be described later in this book. Approaching $b = − 1.0$, further explosions are found.

Stage 7: $b = − 0.975$

The densely dotted region of Figure 4-17 is a large chaotic attractor, an annular chaotic area, d. The frontier, F, of its basin of attraction, includes the inset of the 2-cycle, $\{Q_1, Q_2\}$. The boundary of d is very near F. This figure shows that the critical curves (on the boundary of the former chaotic area) are about to touch (and then to cross) the inset of the 2-cycle.

FIGURE 4-15.

The 11-cyclic chaotic attractor. The permutation of the pieces follows the numbering of Figure 4-14.

a=0.7   b=-0.78

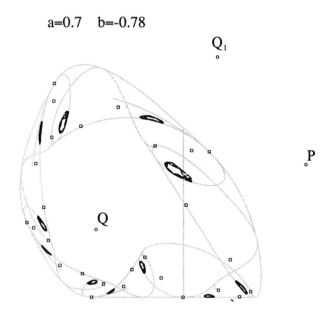

FIGURE 4-16.

a=0.7   b=-0.7989995

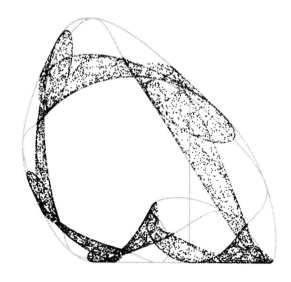

FIGURE 4-17.

a=0.7   b=-0.974

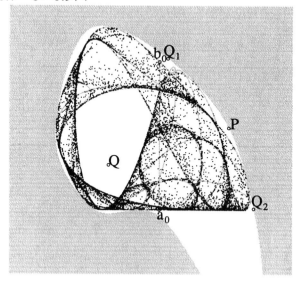

The enlargement of Figure 4-18 shows that, in addition, a contact of the frontier, F, with the boundary on L is about to occur. This contact bifurcation, described in Chapter 7, has the effect of destroying the chaotic attractor, or rather, of transforming it into a chaotic repellor. Now, almost all of the trajectories diverge to infinity, except for a Cantor set surviving inside the former chaotic area.

A rough idea of the basin of infinity, $D(\infty)$, is shown in Figure 4-19 as a black area. The basin of infinity includes infinitely many holes in the former absorbing area, d', only a few of which are shown in this figure. The light area is the basin of attraction of the attractor d. An enlargement is shown in Figure 4-20.

FIGURE 4-18.

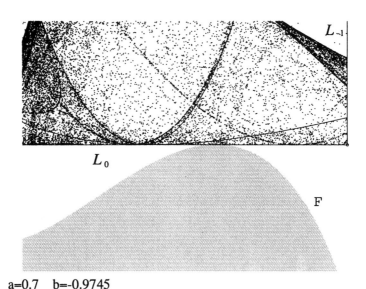

a=0.7   b=-0.9745

FIGURE 4-19.

a=0.7   b=-0.975

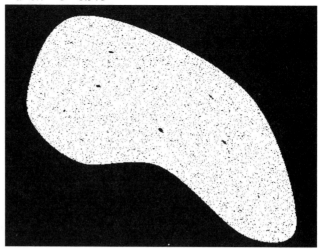

FIGURE 4-20.

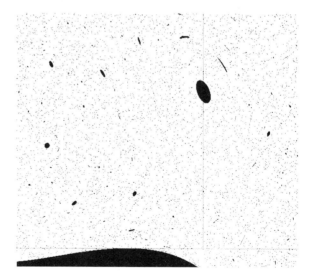

HOLES

5.1 INTRODUCTION

We have already encountered holes in the Case 1 of the first map family treated at the end of the preceding chapter (see Figure 4-17). We now change the parameter a from 0.7 to 1.0, obtaining Case 2 of the first map family, in which the bifurcations involving holes are somewhat clearer. This change eliminates the repelling 2-cycle, $\{Q_1, Q_2\}$. As before, the fixed point Q becomes a repelling focus, but the fixed point P is now a saddle.[1]

The main qualitative features in the portrait of the iterations in this case are: a bounded absorbing area, d', its basin of attraction, $D(d')$, the basin of attraction of infinity, $D(\infty)$, and the boundary between these two basins, F, which consists of the saddle P and its inset.

As the parameter b decreases from zero, the fixed point Q becomes a repelling focus, giving rise to an attractive closed invariant curve, Γ, as before. As Γ crosses L_{-1} there are several bifurcations and an annular absorbing area is obtained, bounded by a finite number of critical arcs, as before. Inside this annular absorbing area, a chaotic area appears.

5.2 EXEMPLARY BIFURCATION SEQUENCE

In this chapter we present a very informative bifurcation sequence, including some new phenomena, ideas and observations.

---

1. See Figures 3-4 to 3-8 to recall the meaning of these terms.

We discuss some of these events now, as b decreases from 0.593 to 0.600. We proceed in eleven stages.

Stage I: $b = -0.59300$

Using Procedure 2 as in the previous chapter, we find an absorbing area d', bounded by seven images of the straight line segment $b_0 a_0$ of L_{-1}. This is the shaded area in Figure 5-1. Note that the boundary of d' includes arcs of L_5 and L_6.

Inside d' there is also an attracting chaotic area, d, as shown in Figure 5-2. Within d' there is an annular absorbing area d'_a, which contains d, and is bounded by the iterates of the line segment $a_0 b_0$ of L_{-1} shown in Figure 5-1. This absorbing area is shown in Figure 5-3, an enlargement is shown in Figure 5-4.

These critical arcs also define the boundary of the chaotic attractor, shown in Figure 5-5 as a densely dotted region. Note that the boundaries of d'_a and d include arcs of L_9 and L_{10}, as shown in Figure 5-4, and in fact the entire boundary of d may be defined by critical arcs.

The basin of attraction $D(d')$ is shown in Figure 5-6, in which the gray region denotes the basin of infinity, $D(\infty)$, The boundary between these two basins, F, is smooth, and consists of the inset of the fixed saddle point, P.

Note the corners of the arc of L_6 on the boundary of d', shown in Figure 5-1. This roughness does not occur for higher values of the bifurcation parameter, b, for which the boundary of d' is smooth. The appearance of this arc gives the first *tongue*, a folding arc of a critical curve, creating roughness of the boundary. This roughness will increase as b continues to decrease, announcing the approach of a *contact bifurcation*.

FIGURE 5-1.

The absorbing area, shaded; and bounded by arcs of critical curves.

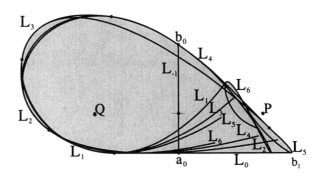

$a=1.0 \quad b=-0.593$

FIGURE 5-2.

The attractor, densely dotted by an actual trajectory.

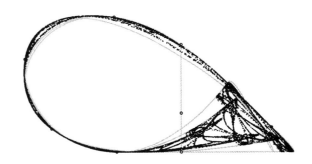

$a=1.0 \quad b=-0.593$

FIGURE 5-3.

With two new segments. The annular absorbing area, shaded, and bounded by arcs of critical curves, the images of these two segments.

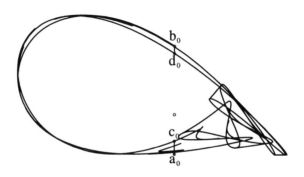

$a=1.0 \quad b=-0.593$

FIGURE 5-4.

An enlargement showing critical curves.

$a=1.0 \quad b=-0.593$

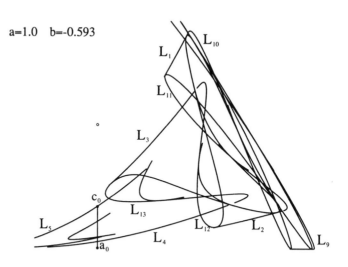

FIGURE 5-5.

An enlargement show-
ing the attractor
bounded by critical
curves.

a=1.0   b=-0.593

L_1
L_{10}
L_{11}
L_3
c_0
L_{13}
L_{12}
L_2
L_5
L_4
L_9
a_0

FIGURE 5-6.

The attractor in its
basin, shaded.

a=1.0   b=-0.593

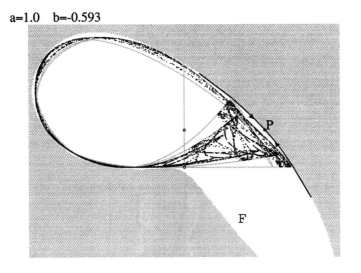

P

F

Stage 2: $b = -0.59495$

This stage immediately precedes the first contact bifurcation, a type of bifurcation treated in more detail later, in Chapter 7. Figures 5-7 and 5-8, with 17 images of the segment $a_0 b_0$, show many tongues in the absorbing area, d'.

The chaotic set, d, contains only part of these tongues, as shown in Figure 5-9. The tongues of the boundary of d' are approaching the inset of the saddle point, P, and thus the boundary of the basin $D(d')$. Note that d' is close to F in Figure 5-9.

Figure 5-10 shows a smaller annular absorbing area, d'_a in d', containing d. Figure 5-11 shows more detail. Figure 5-12 shows that d'_a, and d inside it, are still far from the frontier, F, of the basin of attraction.

Stage 3: $b = -0.594962$

This stage is almost exactly the moment of the first contact bifurcation between the absorbing area, d', and the frontier, F. Figure 5-13 shows that infinitely many tongues on the boundary of d' approach the fixed saddle, P, which lies on F. These tongues belong to images of the segment $a_{-1} a_0$ of L_{-1}. Thus we have a contact between the boundary of the absorbing area, d' and the boundary, F, of its basin, D. Figure 5-14 shows an enlargement near the fixed saddle.

In a further enlargement, Figure 5-15, we see a point of contact, h_0, between the boundary of d' and F. The iterates of this point converge to the saddle, P, as shown in Figure 5-16. And at each of these image points, the images of the two boundaries are tangent; that is, the ends of the tongues are tangent to the inset of the saddle, P.

FIGURE 5-7.

Six critical curves.

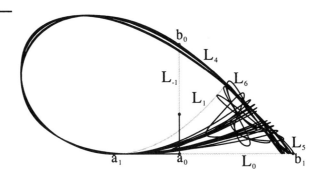

a=1.0   b=-0.59495

FIGURE 5-8.

Seventeen critical
curves, enlarged,
showing tongues.

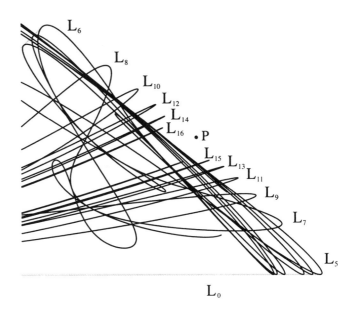

FIGURE 5-9.

The attractor, enlarged.

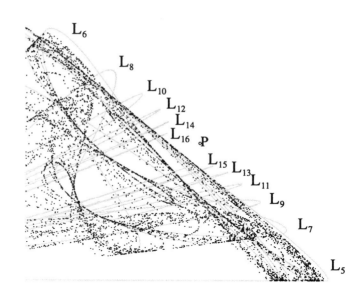

FIGURE 5-10.

The smaller annular
absorbing area, defined
by iterates of a reduced
arc of a critical curve.

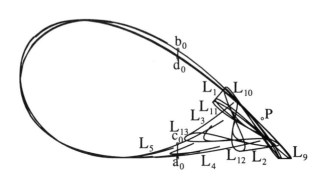

a=1.0   b=-0.59495

FIGURE 5-11.

An enlargement of the shorter critical arcs.

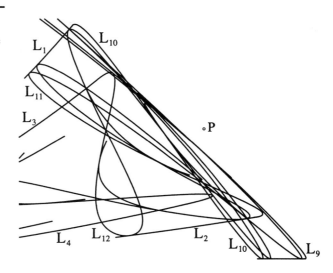

FIGURE 5-12.

A portion of the attractor and its basin, in the reduced annular absorbing area.

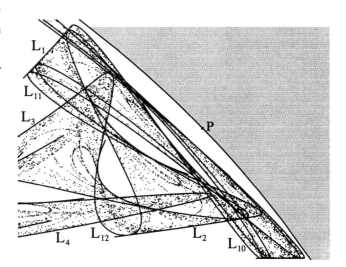

FIGURE 5-13.

The attractor, among
critical arcs.

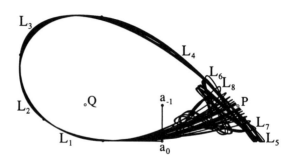

$$a=1.0 \quad b=-0.594962$$

FIGURE 5-14.

An enlargement, show-
ing portions of critical
curves and basin. Note
the tongues tangent to
the inset of *P*.

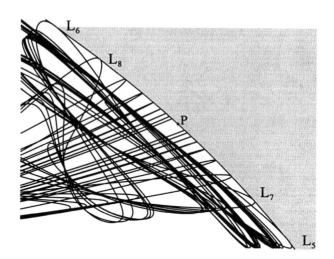

FIGURE 5-15.

Further enlargement,
showing a point of
tangency.

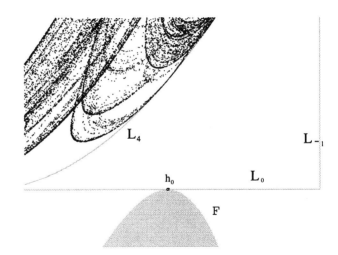

FIGURE 5-16.

Iterates of the point
of tangency converging
to *P*.

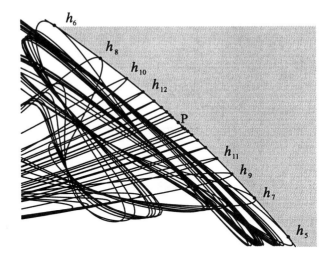

Stage 4: $b = -0.59500$

This stage is immediately after the first contact bifurcation. Figures 5-17 and 5-18, are made of 18 iterates of the segment $a_0 b_0$ of L_{-1}, where b_0 is the intersection of S_4 with L_{-1} (see Figure 5-1). They show that some points of the tongues, having crossed through the inset of P separating the two basins, now are attracted to infinity. Thus, the area defined by Procedure 2 (see 4.3 above) is unbounded; infinite iterations are required to obtain an area which is absorbing. However, a smaller annular absorbing area, d'_a, containing d, may be constructed.

Stage 5: $b = -0.59520$

This stage is also just after the first contact bifurcation. Because the frontier, F, is the inset of the saddle point, P, it is invariant under the inverses of the map, by definition.

Figure 5-19 shows that after the first contact bifurcation, holes (such as those labelled H_{-1}, H^1_{-2}, and H^2_{-2}) appear in the basin of attraction $D(d')$. These holes belong to the basin of infinity.

Note that topologically, the basin of infinity is not connected. It has disjoint pieces, which are holes of the basin of d'. And this basin is not simply connected, as it has holes which belong to the basin of infinity.

This is how the holes appear. After this first contact bifurcation, the frontier F crosses L, creating the sector H_0 bounded by F and L, as shown in the enlargement, Figure 5-20. This sector constitutes a piece of the basin of infinity in the zone Z_2. Since the sector H_0 belongs to the basin of infinity, so too do all of its preimages. One of these, H_{-1}, is shown as a small shaded hole in Figure 5-19. It is in the zone Z_2. (The other, not shown, is in the zone Z_0.) The shaded holes H^1_{-2} and H^2_{-2} are the two first-rank preimages of the hole H_{-1}.

The sector H_0 is bounded by an arc of F_e, and an arc of L having endpoints r_0 and s_0. The first-rank preimage, H_{-1}, of the sector is composed of two areas joined by the arc $r_{-1} s_{-1}$ of L_{-1}. Thus H_{-1} is connected and it is a hole, as shown in Figures 5-19 and 5-20. We regard this as a main hole. All other holes are preimages of a main hole, and they converge to the points Q and Q_{-1}.

We may regard F as the union of F_i and F_e, where F_i consists of the boundaries of all the holes, and F_e is the rest of the boundary of $D(d')$.

For further analysis of bifurcations involving contact of F and L, see BB. We will just describe some of the events in our present context. As b continues to decrease, the holes increase in size.

FIGURE 5-17.

Eighteen iterates of a critical segment.

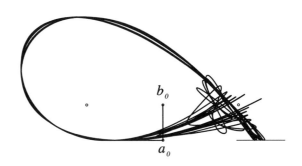

a=1.0   b=-0.595

FIGURE 5-18.

Enlargement showing
critical arcs and basin.
Some tongues now
cross the inset of *P*.

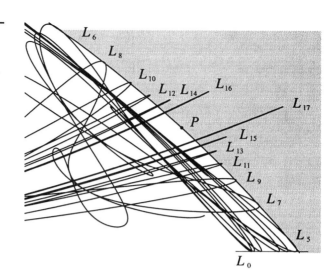

FIGURE 5-19.

The attractor and its
basin.

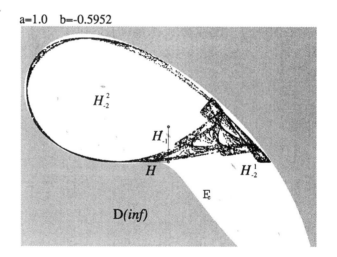

FIGURE 5-20.

Enlargement, showing
the intrusion of the
basin across the critical
line. The domain of
this enlargement is
near the center of
Figure 5-19.

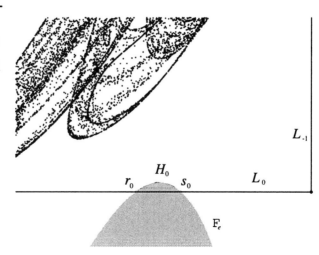

Stage 6: $b = -0.59600$

The preimage of H_{-1} (first-rank preimage of H_0) has two parts, and previously (when b was higher) one of them (denoted by H_{-2}^1) belonged to the region, Z_0, of points having no preimage. But now, as b decreases further, this preimage becomes tangent to L, which is the frontier between Z_0 and Z_2, and crosses through it into Z_2.

This crossing is shown in Figure 5-21. Thus a new set of holes is created, an infinite sequence of preimages of H_{-2}^1 disjoint from our previous system of holes. All these holes, old and new, get mapped eventually into the main hole, H_0.

As b decreases further, the holes increase yet further in size.

Stage 7: $b = -0.59740$

In Figure 5-22 we see the same holes as in the preceding stage, but they are wider. The hole H_{-4}^{11}, which belongs to the region Z_0, is close to the critical curve, L.

See also the enlargement, Figure 5-23.

FIGURE 5-21.

The holes grow larger.

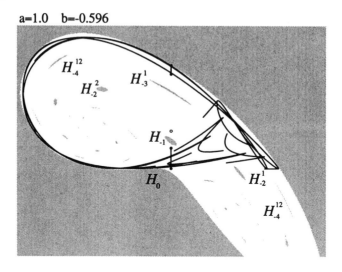

a=1.0 b=-0.596

FIGURE 5-22.

The holes are even
larger.

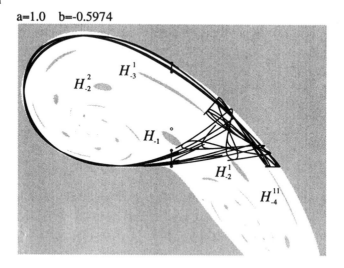

a=1.0 b=-0.5974

CHAOS IN DISCRETE DYNAMICAL SYSTEMS

FIGURE 5-23.

An enlargement.

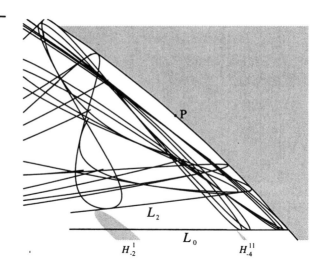

Stage 8: $b = -0.59800$

After further decreases in the parameter b, H_{-4}^{11} now intersects L, as shown in Figure 5-24. Thus, we have passed another contact bifurcation, and we have a new system of holes, based on the hole H_{-4}^{11}.

In the enlargement, Figure 5-25, we see part of H_{-4}^{11} in the zone Z_2, and its image under two iterates of the map lies in the part of H_{-2}^1 above the critical curve, L_2. At the recent contact bifurcation, when H_{-4}^{11} became tangent to L, H_{-2}^1 became tangent to L_2, and H_{-1} became tangent to L_3. Thus this contact bifurcation changed the topology (that is, the density of holes) of both the basin $D(d'_a)$, and its annular absorbing area, d'_a.

In Figure 5-25, we see that the critical arcs of L_{10}, L_{11}, and so on, cross the frontier, F_e. Before the recent contact bifurcation, these arcs defined the boundary of an absorbing area. Notice also in Figure 5-25 the hole $\overline{H}_{-k}$, a preimage of the new hole $\overline{H}_{-4}^{11}$ which approaches the critical curve L from above, that is, from the zone Z_2. Now a new absorbing area exists, and in Figure 5-26 we see that its boundary includes arcs of L_{11} and L_{12}, without contacts with F_e.

Stage 9: $b = -0.59820$

In Figure 5-27, the hole $\overline{H}_{-k}$ becomes tangent to L.

HOLES   **75**

FIGURE 5-24.

A new system of holes perforates the basin of our attractor.

a=1.0   b=-0.598

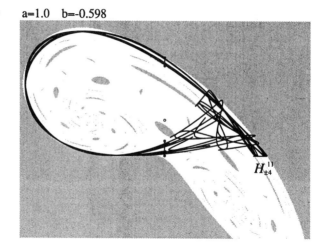

$H_{\pm4}^{11}$

FIGURE 5-25.

Darker shading indicates the new tongues crossing into the basin of infinity.

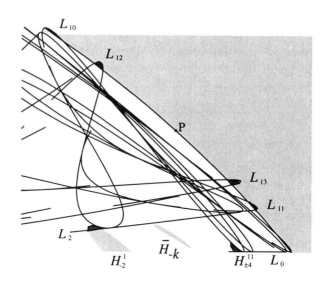

L_{10}

L_{12}

P

L_{13}

L_{11}

L_2

$\bar{H}_{-k}$

H_{-2}^{1}

$H_{\pm4}^{11}$ L_0

FIGURE 5-26.

The new absorbing area, bounded by these critical curves.

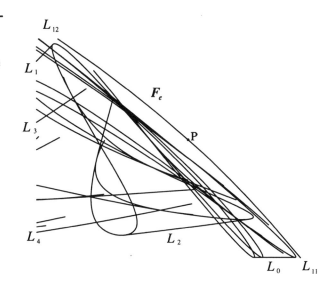

FIGURE 5-27.

The hole has descended to L from above.

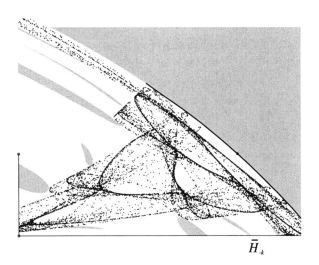

Stage 10: $b = -0.59824$

In Figure 5-28 the hole $\overline{H}_{-k}$ crosses L. In this third contact bifurcation, we see holes rejoined, the inverse of the second contact bifurcation discussed in Stage 9. What had been two distinct holes (and their preimages) are now reunited, connected by a segment (and its preimages) in L_{-1}. This is marked "reunion" in Figure 5-28, compare Figure 5-24.

In the enlargements, Figures 5-29 and 5-30, we can see images of critical arcs defining the boundary of the annular absorbing area, d_a', and the chaotic area, d. Notice that the hole H_{-f} is in zone Z_0, but is very close to L, which belongs to the boundaries of both d_a' and d. With further decreases of b, this hole makes contact with L. It may be established that such a contact will be the next contact bifurcation, but its effect will be different from the preceding ones.

FIGURE 5-28.

Some holes have now rejoined.

a=1.0   b=-0.59824

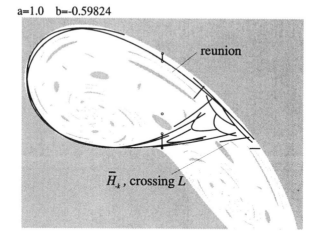

reunion

$\overline{H}_{-k}$, crossing L

FIGURE 5-29.

Enlargement, show-
ing the new holes
below L in darker
shading.

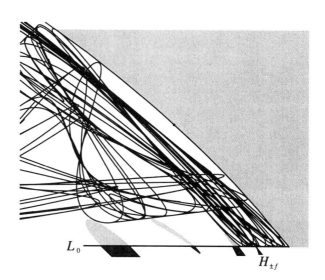

FIGURE 5-30.

Enlargement, showing
the attractor and basin
with holes.

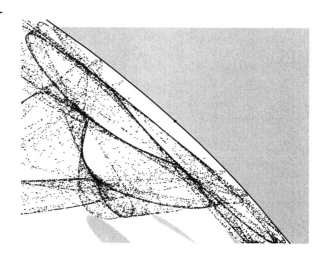

Stage 11: $b = -0.598727$

This is the stage of contact predicted in Stage 10. The hole, H_{-f}, is tangent to L. The boundary F becomes tangent to the boundary of d at infinitely many points. These points comprise the trajectory of the point k_0 shown in Figure 5-34. See Figure 5-31, and its enlargements, Figures 5-32 to 5-36.

The boundary of d has contact with the hole H_{-f}, as well as with all its images up to the main hole, and the sector H_0 and its images. These create the tangency of infinitely many tongues of d in the inset of P, as shown in Figure 5-36. This is an example of *homoclinic tangency*: the tangency of the outset of the saddle point P to the inset of P. As the trajectory of this point of tangency tends to P in both future and past iterations, it is same-tending, or homoclinic, in the language of Poincaré.

The rank 1 preimage of the point k_0 is a point k_{-1} of L_{-1} within the chaotic area d (see Figure 5-35). This tangency of the attractor and its basin boundary will cause an explosion of holes inside the chaotic area. We leave this to the interested reader to explore using the software ENDO, available on the companion CD-ROM.

The difference between this third contact bifurcation and those proceeding is that we have here a contact between the boundary of d, a chaotic area, and the boundary, F, of its basin. This contact causes the destruction of the chaotic area d, which is changed from an attractor to a repellor. The hole $H_{-(f+1)}$ (the preimage of H_{-f}) possesses an arborescent sequence of preimages inside the chaotic area, d, leaving a chaotic repellor. Nearby trajectories are now attracted to other attractors, at infinity.

FIGURE 5-31.

The third contact bifurcation

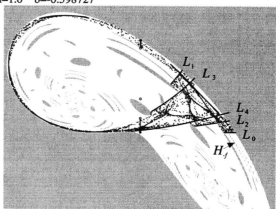

FIGURE 5-32.

Enlargement of a rectangle near the center of Figure 5-31.

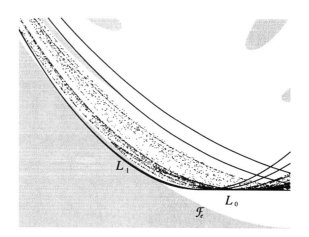

FIGURE 5-33.

The basic hole.

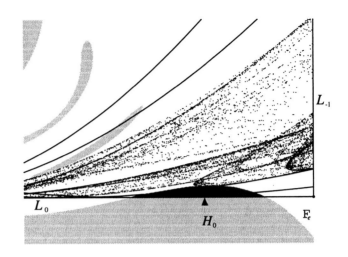

FIGURE 5-34.

The point of contact.

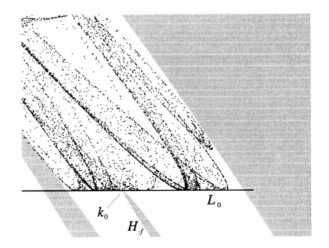

FIGURE 5-35.

A preimage of the
point of contact.

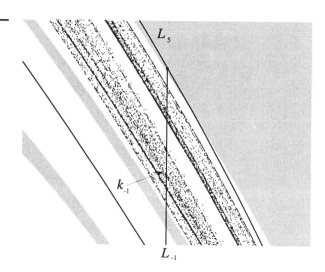

FIGURE 5-36.

Many contacts of the
boundary of the cha-
otic area and the
boundary of its basin.

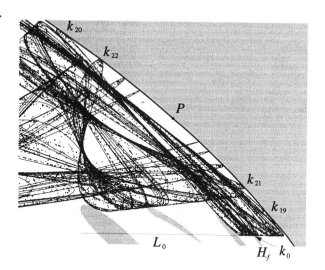

CHAPTER 6

FRACTAL BOUNDARIES

Fractal boundaries are very important in the applications of discrete dynamics. First we describe in greater detail the basic features of contact bifurcations, which we have encountered already in Chapter 5, with a sequence of hand drawings. Then we will go on to an exemplary bifurcation sequence with computer graphics, in which the fractal implications of these contact events for the boundaries become clear.

6.1 CONTACT BIFURCATION CONCEPTS

We now introduce the simplest kinds of contact bifurcations, which are global bifurcations in the sense that the topology of the basins changes.

6.1.1 The loss of simple connectivity

First we show the transition of a basin D from simple to multiple connectedness due to the appearance of holes in D. In Figure 6-1, D (shown shaded) is simply connected; there are no holes. In Figure 6-2, a contact occurs at the point h_0 between $\mathcal{F}$, the frontier of D, and L (also denoted L_0 in the figures), the critical curve of rank 1. The rank 1 preimage of h_0 is a point, h_{-1}, of L_{-1} belonging to D. This point and all of its preimages are germs of holes; they become holes after the crossing of $\mathcal{F}$ through L.

In Figure 6-3, after the crossing of $\mathcal{F}$ and L, the sector H_0 has a rank 1 preimage in the hole H_{-1}, and this hole no longer belongs to D, nor do its preimages.

FIGURE 6-1.

The shaded region is
the basin, D. The two
fixed points are
labelled P and Q.

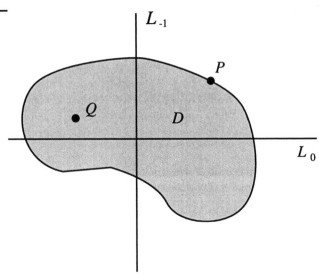

FIGURE 6-2.

A contact occurs at a
single point.

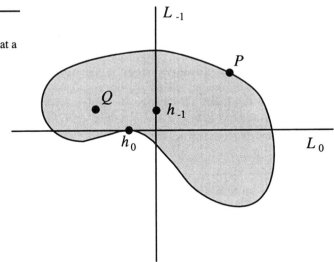

FIGURE 6-3.

The two contact points border the main hole, shown here with its preimages (holes) of ranks 1, 2, and 3.

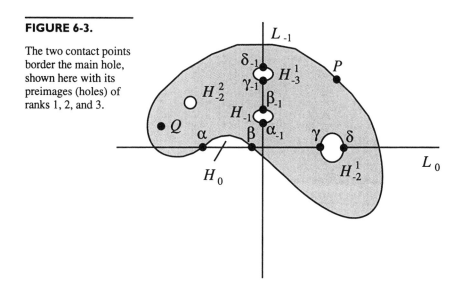

6.1.2   The loss of connectedness

Next, we show the transition of a basin D from connectedness to disconnectedness; that is, from one to many pieces.

In Figure 6-4, the basin D is connected. In Figure 6-5, a contact between $\mathcal{F}$ and L occurs at the point h_0 of L. Its rank 1 preimage h_{-1} in L_{-1}, and all the preimages of h_{-1}, are outside of D. They are germs of components of D (*islands*), and become islands after the crossing of $\mathcal{F}$ through L.

In Figure 6-6, after the crossing of $\mathcal{F}$ and L, the sector (*headland*) Δ_0 has a rank 1 preimage in the island D_{-1}. This island belongs to D, as do its preimages. In this case the component of the basin D containing the attractor is called the *immediate basin, D_0*.

FIGURE 6-4.

The shaded region is the basin, D.

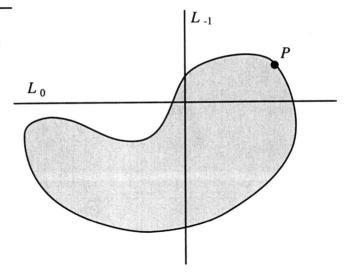

FIGURE 6-5.

Contact is made at a single point.

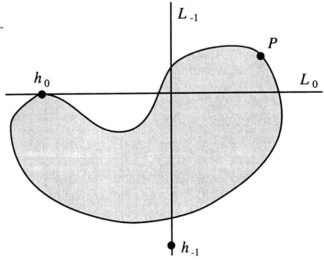

CHAOS IN DISCRETE DYNAMICAL SYSTEMS

FIGURE 6-6.

The two points of contact border a headland, shown here darkly shaded. One preimage (island) is also shown.

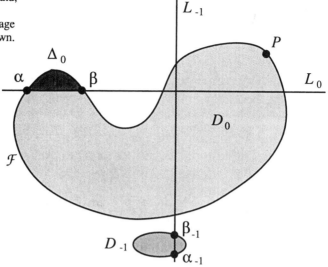

6.1.3   Distinguishing two types of contact bifurcation

Now we must assume the notions of open and closed subsets of the plane, as presented in advanced calculus, for example. Recall that the *closure* of a subset of the plane, S, is the smallest closed set containing S, for which we use the notations $cl(S) = \bar{S}$. The *boundary* of a subset S of the plane is defined as the common points of the closure of S and the closure of the complement of S; that is $bd(S) = \partial S = \bar{S} \cap cl(R^2 \backslash S)$.

The two types of contact bifurcation shown in 6.1.1 and 6.1.2 are similar, but are distinguished by the behavior of the point h_{-1} of L_{-1}. In these two cases, Figure 6-2 and Figure 6-5, we may see two simple rules.

- If the point h_{-1} belongs to the portion of L_{-1} inside $cl(D)$, then holes will appear.

- On the other hand, if h_{-1} belongs to the part of L_{-1} outside of $cl(D)$, then islands will appear.

Note: A basin is an open set, that is, every point of D has a neighborhood contained entirely within D. Also, a basin is backwards invariant, that is, all preimages of a point of D belong to D, and in fact, $T^{-1}(D) = D$. Generally, the boundary of a basin, $\mathcal{F} = bd(D)$, is backwards invariant; but there are exceptions, and Figures 6-2 and 6-5 both show this invariance violated, if $\mathcal{F}$ is considered as the boundary of the unbounded basin surrounding D in Figure 6-2 and the boundary of the bounded basin D in Figure 6-5. As we are interested in describing the bifurcation of the bounded basin we may use the two following bifurcation rules:

- When $T^{-1}(\mathcal{F}) \setminus \mathcal{F}$ belongs to $cl(D)$ at a contact bifurcation, then holes will appear.

- When $T^{-1}(\mathcal{F}) \setminus \mathcal{F}$ does not belong to $cl(D)$, then islands will appear.

Notice that the transition from $T^{-1}(\mathcal{F}) = \mathcal{F}$ to $T^{-1}(\mathcal{F}) \neq \mathcal{F}$ can occur only if for some point of $\mathcal{F}$, the number of preimages increases. Recall that the preimages of a point x can increase only by the crossing of x through L. Thus, these bifurcations of a basin must involve a contact between $\mathcal{F}$ and L.

If h_0 in L is a point of contact between $\mathcal{F}$ and L, and h_{-1} is its unique rank 1 preimage belonging to L_{-1}, then the iterated preimages (that is, of rank 1, rank 2, and so on) of h_{-1} may be finite in number (ending in a point with no preimages), or infinite within a finite number of sequences (due to some points having a finite number of preimages of rank 1), or even infinite with an infinite number of sequences (due to some point having an infinite number of preimages of rank 1), and chaotic.

Warning: The simple rules described above are only sufficient conditions for the appearance of holes or islands. We will see soon that contact bifurcations causing the appearance of holes or islands may occur even without exception to the general situation, $T^{-1}(\mathcal{F}) = \mathcal{F}$. This situation arises in bifurcations involving the reunion of holes or islands, as described in the following figures.

6.1.4   Further contacts

The first three drawings, Figures 6-7, 6-8, and 6-9, are the same as Figures 6-4, 6-5, and 6-6. After Figure 6-9, the components of the disconnected basin, which are given by the preimages of the headland, Δ_0, increase and come together, touching in Figure 6-10. Their reunion after this bifurcation is shown in Figure 6-11.

The points of the arc of the frontier $\mathcal{F}_0$ of D_0 external to the boundary of the headland F_i have preimages which belong to the immediate basin D_0 in Figure 6-9.The points of F_i have preimages in the boundaries of the other components of the total basin, D.

In Figure 6-10, the boundary of the immediate basin and the boundary of the main component, D_{-1}, have contact at the point k_{-1}, which belongs to $\mathcal{F}_0$, to the boundary of D_{-1}, and to the critical curve L_{-1}. All preimages previously disjoint now have a contact point in the preimages of k_{-1}.

In Figure 6-11, after this contact bifurcation, all of the former components have been reunited.

Note: Beginning with the disconnected basin of Figure 6-9, a bifurcation sequence might proceed either through Figure 6-8 to Figure 6-7, with the components decreasing and disappearing, or through Figure 6-10 to Figure 6-11, with the components increasing and reuniting. A similar sequence might occur with holes (islands) as follows.

FIGURE 6-7.

Compare with
Figure 6-4.

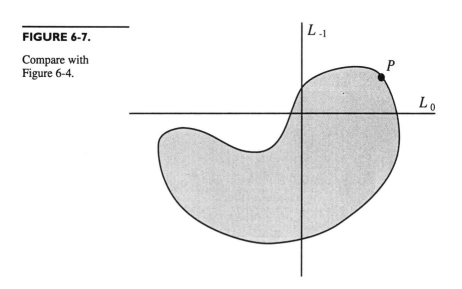

FIGURE 6-8.

Compare with
Figure 6-5.

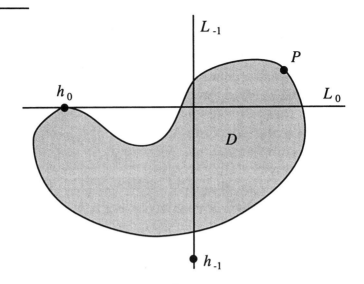

FIGURE 6-9.

Compare with Figure
6-6. Here is a head-
land, darkly shaded,
and its preimage, an
island.

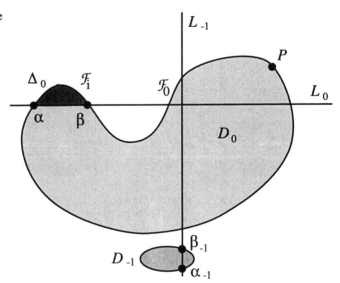

FIGURE 6-10.

Here is a new point of
contact at one end of
the headland, and
another on the island.

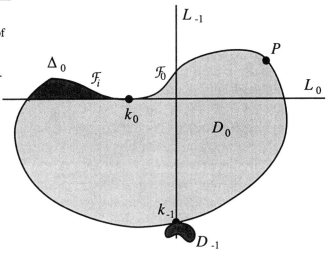

FIGURE 6-11.

The island has rejoined
the mainland.

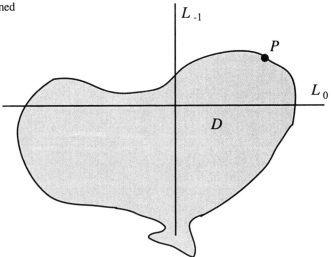

6.1.5   An alternative contact scenario

Figures 6-12, 6-13, and 6-14 are identical to Figures 6-1, 6-2, and 6-3 of this chapter. As the bifurcation sequence progresses, the holes inside the basin D increase and get closer together.

In Figure 6-14, the preimages of the points in $\mathcal{F}_e$ belong to the external boundary, while the preimages of the points of $\mathcal{F}_i$ (the boundary of H_0) belong to the boundaries of the internal holes, called internal boundaries. In Figure 6-15, all the holes which are preimages of the sector H_0 meet in another contact bifurcation. The external frontier has met the internal frontier, due to the point of tangency with L, k_0. In particular, the main hole, H_{-1}, has made contact with the external frontier, $\mathcal{F}_e$, at the point k_{-1}, which belongs to $\mathcal{F}_e$, to the boundary of H_{-1}, and to L_{-1}. In Figure 6-16, after the bifurcation, all the holes have disappeared due to the contact in Figure 6-15, by opening to the sea outside.

Note: Beginning with the multiply-connected basin of Figure 6-14, a bifurcation sequence might proceed either through Figure 6-13 to Figure 6-12, with the holes decreasing and disappearing, or through Figure 6-15 to Figure 6-16, with the holes increasing and reuniting. Also note that the sequence of Figures 6-12 to 6-16, viewed from the complementary basin, $D(\infty)$, is analogous to the sequence of Figures 6-7 to 6-11.

FIGURE 6-12.

Like Figure 6-1.

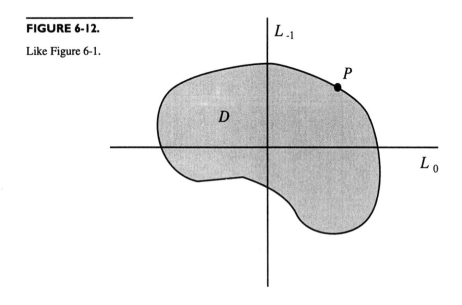

FIGURE 6-13.

FIGURE 6-13.

Like Figure 6-2.

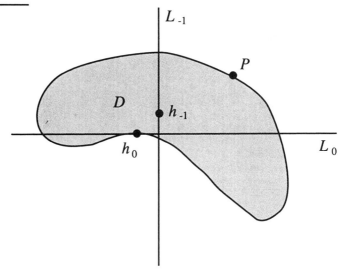

FIGURE 6-14.

Like Figure 6-3.

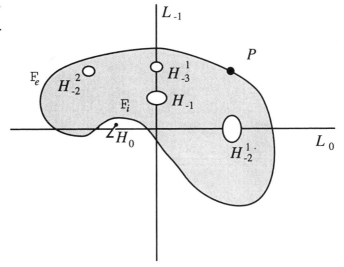

FIGURE 6-15.

A contact of holes.

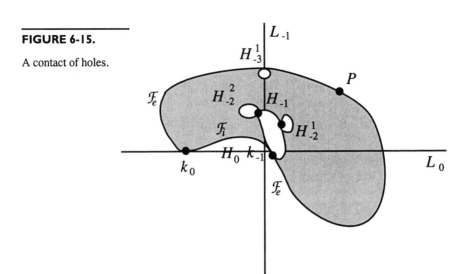

FIGURE 6-16.

The holes join to the sea.

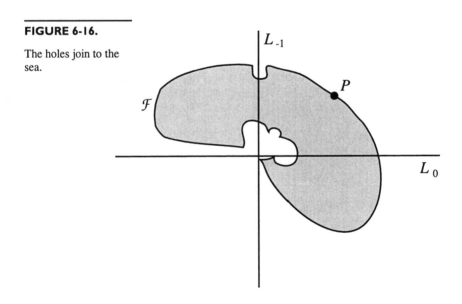

6.2   EXEMPLARY BIFURCATION SEQUENCE

Here again we use the first family of quadratic maps, this time with $a = -1.5$. As before, this map family has fixed points P and Q, and a repelling 2-cycle, $\{Q_1, Q_2\}$.

As the bifurcation parameter, b, decreases from zero, the fixed point Q changes from an attractor to a repelling focus, giving rise to an attractive invariant closed curve, Γ, which crosses L_{-1}, undergoes several bifurcations, as a result of which there appears an annular absorbing area containing a chaotic attractor.

The fixed point P is now a repelling node of a special kind (which we call a *cusp point*; see Appendix A3.2) on the frontier $\mathcal{F}$, which is the boundary separating two basins, the basin D of an absorbing area d', and its complement, $D(\infty)$, which is the basin of infinity, that is, the set of all points whose trajectories recede to infinity.

In this family of maps, however, new kinds of local and global bifurcations take place on the frontier $\mathcal{F}$. The 2-cycle becomes a repelling node, and undergoes a Myrberg sequence of period-doubling bifurcations, which create an infinity of cycles on F. Also, the cusp point, P, creates a new kind of global bifurcation due to contacts between $\mathcal{F}$ and L. In this situation, the quality of the fractal structure of $\mathcal{F}$ changes from soft to hard, as we shall see. We are going to decrease b from -1.5 to -2.115, in 12 stages.

Stage 1: $b = -1.50000$

The basin D begins life as the basin of attraction of the fixed point Q, while it is still attracting. Here, as shown in Figure 6-17, there is a first contact between $\mathcal{F}$, the boundary of D, and L. The rank 1 preimage of P, the point P_{-1}, crosses L, moving from zone Z_0 to Z_2. After this contact bifurcation, the basin D is disconnected.

FIGURE 6-17.

Contact bifurcation.

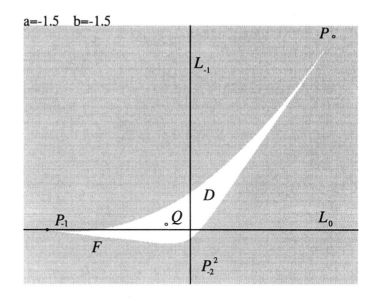

$a=-1.5 \quad b=-1.5$

Stage 2: $b = -1.59000$

In Figure 6-18, after contact, we see that D has two components, the immediate basin, D_0, and a component belonging to the zone Z_0, D_{-1}, and thus having no preimages. This first contact bifurcation has disconnected the basin, D, and is of one of the types described in Figure 6-2, or equivalently, Figure 6-9. Note that the fixed point Q is now repelling, and has emitted an attractive invariant closed curve, Γ. Recall that the regions R_1 and R_2 are folded onto the zone Z_2 by the map.

FIGURE 6-18.

After contact. Note the small island.

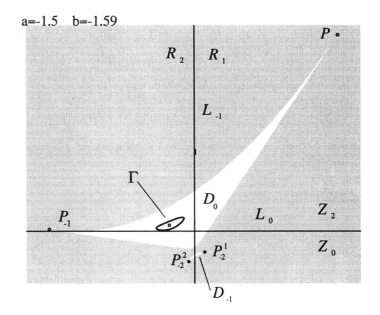

a=-1.5   b=-1.59

Stage 3: $b = -1.60100$

The second contact bifurcation occurs at this value. As shown in Figure 6-19, the two components of D are reunited, and D again becomes connected. The closures of D_0 and D_{-1} meet at the point k_{-1} of L_{-1}. Note that an arc of the frontier of D_0 is tangent to L at k_0.

FIGURE 6-19.

Second contact.

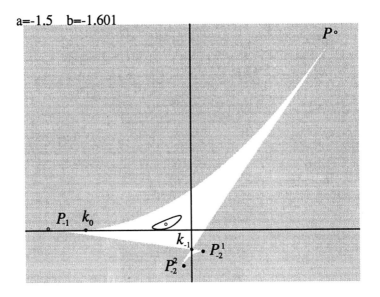

CHAOS IN DISCRETE DYNAMICAL SYSTEMS

Stage 4: $b = -1.75000$

In Figure 6-20, after the second contact bifurcation, the basin D is again connected, due to the reunion of the two components. This is an instance of the contact bifurcation described in Figures 6-9, 6-10, and 6-11.

These first four stages comprise exactly the sequence of Figures 6-7 to 6-11 of Section 6.1. The next stages comprise the sequence of Figures 6-12 to 6-16 of Section 6.1.

FIGURE 6-20.

After the second contact.

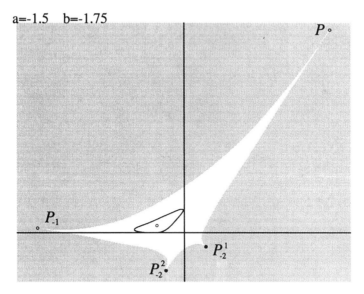

a=-1.5   b=-1.75

Stage 5: $b = -1.95021$

In Figure 6-21, we see that the boundary of D, coming from Z_0, has made contact with L at h_0, exactly as in Figure 6-13.

FIGURE 6-21.

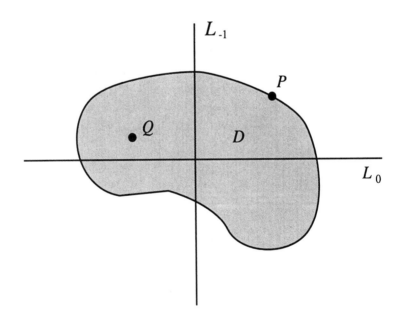

CHAOS IN DISCRETE DYNAMICAL SYSTEMS

Stage 6: $b = -1.95500$

In Figure 6-22, we see the main hole, H_{-1}. Infinitely many sequences of holes also exist. At lower values of b, these holes become wider, and there are sequences of bifurcations like those already observed in the map family of Chapter 5. That is, holes crossing from Z_0, through L, into Z_2 cause other explosions of sequences of preimages. And holes crossing from Z_2, through L, into Z_0 cause reunions of holes.

Remark. This situation differs from that of the preceding chapter, in which the holes, preimages of H_{-1}, have only two accumulation points (the repelling focus and its rank 1 preimage.) But here, the preimages of H_{-1} accumulate on several repelling cycles belonging to $\mathcal{F}$.

FIGURE 6-22.

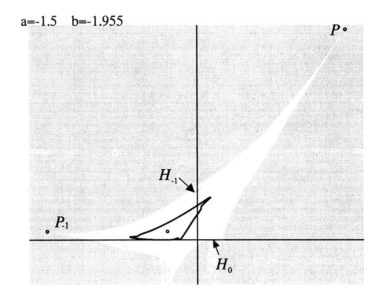

a=-1.5 b=-1.955

P

H_{-1}

P_{-1}

H_0

Stage 7: $b = -1.98000$

We will now see that the hole sequences have a self-similar structure: in successive enlargements, similar structures will be observed. In Figure 6-23, we see the two basins, D and $D(\infty)$, in a large domain of the plane. The hole H_{-1} and its two preimages of rank 1 are indicated. Note, for future reference, the rectangle drawn around H_{-1}. Some of the cusp points, the fixed point P and its pre-images, are indicated also.

Figure 6-24 is an enlargement of the small rectangle drawn in Figure 6-23. Figure 6-25 to 6-27 are successive enlargements of the indicated rectangles, the self-similarity of the holes is evident. We call this structure a *weak fractal*. The frontier, $\mathcal{F}$, has a weak fractal structure caused by the accumulation on the interior boundary, $\mathcal{F}_i$, of preimages of the main hole, H_{-1}. The fractal structure is weak in the sense that there are only a finite number of cusp points on the frontier. This will change in the next bifurcation. As b decreases further, all the holes approach each other, and all approach the external boundary, $\mathcal{F}_e$.

Figure 6-28 is an instance of the bifurcation shown in Figure 6-15, in which the holes meet each other and the frontier.

FIGURE 6-23.

The big picture.

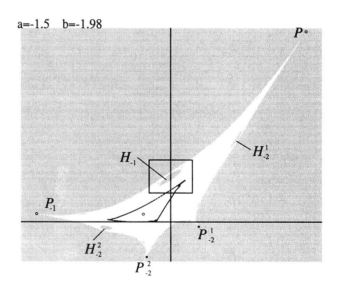

CHAOS IN DISCRETE DYNAMICAL SYSTEMS

FIGURE 6-24.

First enlargement.

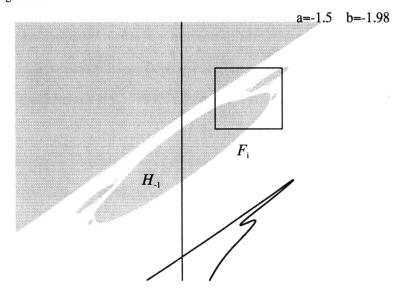

a=-1.5   b=-1.98

F_i

H_{-1}

FIGURE 6-25.

Second enlargement.

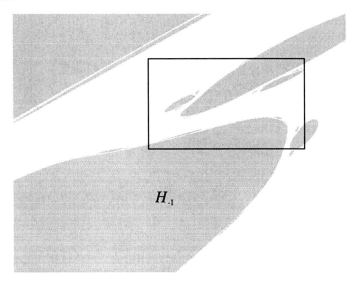

H_{-1}

FIGURE 6-26.

Third enlargement.

a=-1.5   b=-1.98

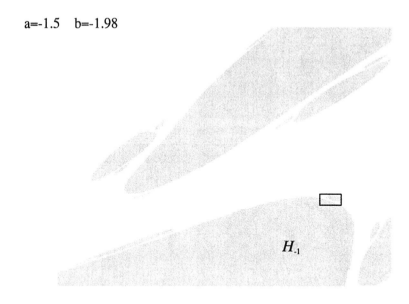

H_{-1}

FIGURE 6-27.

Fourth and final
enlargement.

a=-1.5   b=-1.98

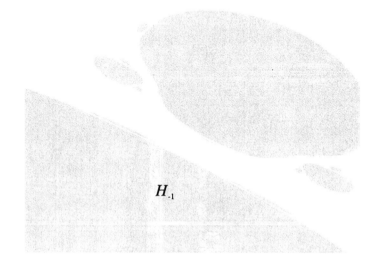

H_{-1}

FIGURE 6-28.

Note the new hole,
darkly shaded.

a=-1.5   b=-2.09

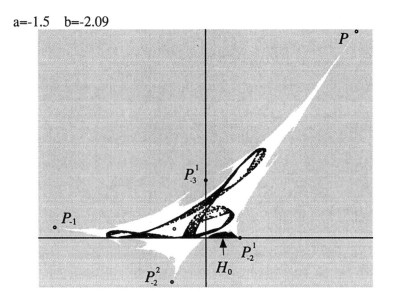

Stage 8: b = − 2.09000

The attractor has exploded.

Stage 9: b = − 2.10000

Figure 6-29 shows that the cusp point, P^1_{-2}, has moved from zone Z_0, through L, into Z_2. All the holes which are preimages of the main hole, H_{-1}, have disappeared. They are now connected to the basin of infinity, $D(\infty)$. An explosion of arborescent sequences of preimages of the cusp point P has created an infinity of cusp points on the external frontier $\mathcal{F}_e$, which now has a strong fractal structure: now there are infinitely many cusps.

Here is a mechanism which might be involved in this bifurcation: Some cusps on the frontier near L may leave headlands, preimages of which are islands inside former holes. An example (before, during,. and after) is shown in Figure 6-30, in which a portion the basin, D, is shown shaded. In this hypothetical situation, the total basin, D, is not connected.

However, in our case the basin D is simply connected, that is, has no holes. This is shown in Figure 6-31, in which the basin D is white and the basin of infinity, $D(\infty)$, is shaded. Note that the black chaotic area, which is multiply connected, and the absorbing area to which it belongs, are near the frontier, $\mathcal{F}$, which separates the two basins. The hole $W(Q)$ contains the repelling fixed point, Q. The three holes labelled W_1, W_2, and W_3, surround a repelling 3-cycle.

It is clear that there is a simply connected absorbing area, d', containing the chaotic attractor, within D. The point $(0, -1.99)$ in L_1 and the point $(0, -1.2)$ in L_2 are the endpoints of a line segment in L_{-1}, S, which is shown in Figure 6-31. The boundary of d' is obtained by a few images of this straight line segment, S.

Finally, we see that we are approaching another contact bifurcation, in which F and L will touch at a point h_0, which belongs also to the boundary of d'. Because the frontier $\mathcal{F}$ is a fractal, or fuzzy, it will be difficult to determine the exact moment of this contact. Figure 6-32 is an enlargement of the lower left rectangle in Figure 6-31, while Figure 6-33 is an enlargement of the rectangle on the right of Figure 6-31. These two enlargements clearly show the approaching contact of $\mathcal{F}$ and L.

The contact of $\mathcal{F}$ and L may be detected by examining the successive images of the segment, S. As long as they belong to d', the bifurcation has not yet occurred.

Stage 10: $b = -2.10300$

Here at last is the contact between $\mathcal{F}$ and L. Figure 6-34 shows the arc of L which is involved in the contact at the point h_0. It is part of the boundary of the image of W_1, and is the image of a small segment of L_{-1} inside W_1. This segment contains the point h_{-1}. The main hole, H_{-1}, will appear in W_1.

FIGURE 6-29.

A cusp point cross-
ing L, the
horizontal line.

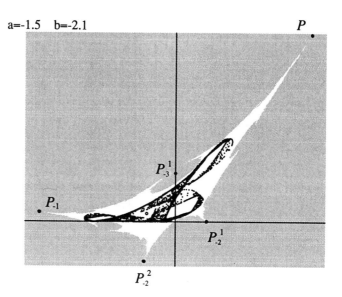

$a=-1.5$ $b=-2.1$

P

P_{-3}^1

P_{-1}

P_{-2}^1

P_{-2}^2

FIGURE 6-30.

Before, during, and
after the hypothetical
bifurcation. The
cusp point moves up
through L, while at
the same time, a hole
merges with the
headland.

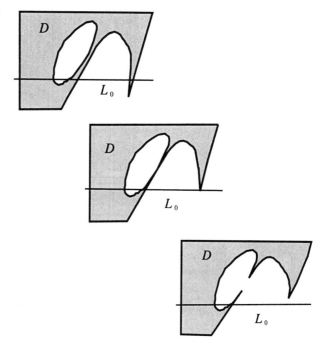

D

L_0

D

L_0

D

L_0

FIGURE 6-31.

The holes, the segment S, and two windows to be enlarged. In these windows, a contact event like that of the preceding figure (or some close relative) is occurring.

a=-1.5   b=-2.1

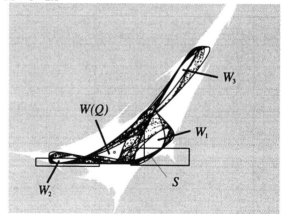

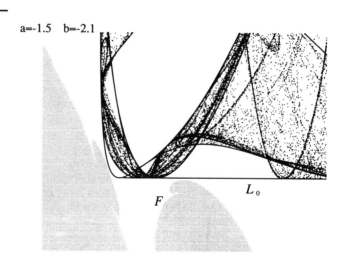

FIGURE 6-32.

a=-1.5   b=-2.1

CHAOS IN DISCRETE DYNAMICAL SYSTEMS

FIGURE 6-33.

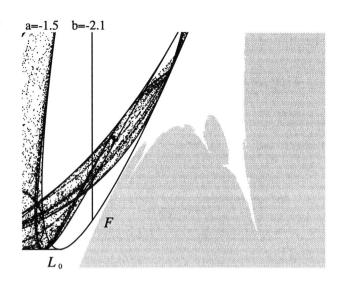

a=-1.5   b=-2.1

F

L_0

FIGURE 6-34.

a=-1.5   b=-2.1024

The frontier F approaching the horizontal segment of L in the lower left. Its preimage is the short vertical segment in the white hole on the right.

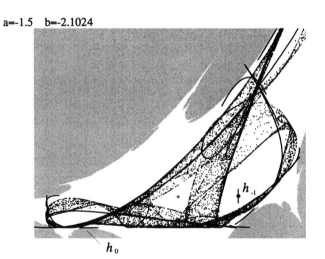

h_{-1}

h_0

Stage 11: $b = -2.11000$

Figure 6-35 gives an overall view of the two basins, after the bifurcation at Stage 10. Figures 6-36 and 6-37 are enlargements of two different rectangular areas, showing holes inside the regions W_2 and W_1, respectively. Figure 6-38, the enlargement of the third rectangle (upper) shown in Figure 6-35, shows the fractal structure of the holes: close to the boundary, there are more and more holes.

An invariant chaotic area, bounded by a finite number of critical arcs, still exists. As b decreases further, however, the holes become wider and approach the chaotic area. Somewhere, a hole is approaching L near a critical arc of the boundary of the chaotic area.

FIGURE 6-35.

The big picture.
Compare with
Figure 6-31.

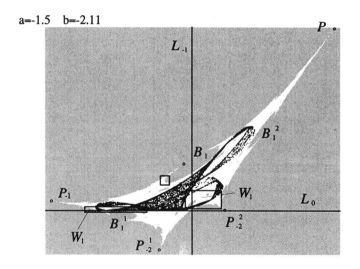

FIGURE 6-36.

Enlargement of the lower left rectangle.

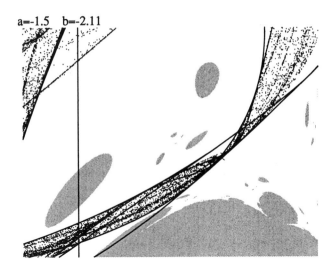

a=-1.5   b=-2.11

FIGURE 6-37.

Enlargement of the lower right rectangle.

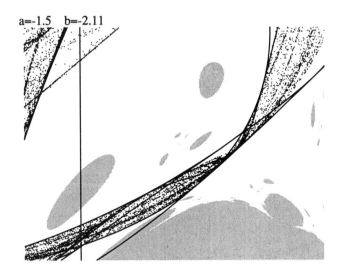

a=-1.5   b=-2.11

FIGURE 6-38.

a=-1.5   b=-2.11

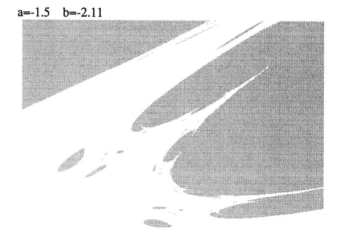

Stage 12: b = – 2.11300

In Figure 6-39, tongues of the chaotic area have appeared which cross the boundary of the old chaotic area. A bifurcation is approaching, in which these tongues (for example, L_{12} and L_{13}) will become tangent to and then cross the frontier $\mathcal{F}$. This occurs at about $b = -2.11400$.

This is the final contact bifurcation in our sequence, as now the former chaotic area has been pierced with holes and destroyed. That is, it has lost invariance and attractivity.

FIGURE 6-39.   a=-1.5   b=-2.113

CHAPTER 7

CHAOTIC CONTACT BIFURCATIONS

Chaotic contact bifurcations involve a chaotic attractor. This is the pinnacle of our subject. Here we proceed with a 1D introduction, and a 2D introduction, before analyzing the exemplary bifurcation sequence.

7.1 BIFURCATIONS IN ONE DIMENSION

Recall from Chapter 3 that the basic critical curve L (as defined in 3.2) belongs to the frontier of zones having different numbers of preimages, for example, between Z_1 and Z_3. These concepts were introduced in the one-dimensional case in 2.2.

We now turn to another context, the *chaotic contact bifurcation*, or *CCB*. Critical points are fundamental to an understanding of these global events, and this understanding will be very useful when we come to consider CCBs in the two-dimensional case.

In one-dimensional iterations, the transition to chaos (a bifurcation sequence in which a chaotic attractor is created out of the blue sky, or a periodic attractor becomes chaotic) has been a primary concern since the pioneering works of Myrberg in 1958 (see Appendices 5 and 6.) The role of the critical points in these transitions has been explored by Mira since 1975. His analysis of the box-within-a-box bifurcation structure is described in M1. The connection between bifurcations due to the critical points and the homoclinic bifurcations of the repelling cycles has been presented in (Gardini, 1994). (Homoclinic points were discussed in Chapter 5.)

This latter is the subject of this chapter. We are going to illustrate two kinds of CCB. These examples are also homoclinic bifurcations. Both concern a chaotic attractor having several pieces — intervals in the one-dimensional case — which are permuted cyclically by the map. These are called *cyclical chaotic attractors*.

Warning: We use the term *chaotic attractor*, loosely, for a situation revealed experimentally. We can never be sure that a trajectory which appears chaotic is a true chaotic attractor, or just a very long chaotic transient.

In a *CCB of the first kind* a cyclical chaotic attractor explodes into a single, larger chaotic attractor. In a *CCB of the second kind*, a $2k$-cyclic chaotic attractor is transformed into a k-cyclic chaotic attractor as pieces merge pairwise.

Our examples will all occur in the quadratic family of maps of the real line studied by Myrberg, $f(x) = x^2 - b$. We will discuss the dynamics as b increases in the interval $[-2, 2]$. We begin with $b = 1.0$ and increase to about $b = 1.8$, in nine stages.

The graph of a Myrberg map is a parabola in standard position, except for the vertical displacement by $-b$. As the bifurcation parameter b increases, the parabola descends. As described in Chapter 2, two fixed points appear in a fold bifurcation as b increases through the value 0.25. Generally, a fold bifurcation is the opening of a *box of the first kind*, in the language of Mira. A *box* is an interval in the one-dimensional space of the parameter, b.

Stage 1: $b = 1.0$

Figure 7-1 shows the main qualitative features of the function f. The two fixed points are shown by the intersection of the graph of f and the diagonal line. Although these points are really located in the domain of the map (the horizontal axis in this figure), we show them above the horizontal axis, on the graph of the function or on the diagonal, which are the same in this case. This convention is very useful, and will be followed throughout this chapter.

The critical point (value of x at which $f(x)$ achieves its minimum) is the origin, 0, and is denoted c_{-1} here. The critical value $f(c_{-1})$ is indicated by c. Of course, this is just the critical value $-b$. Again, both are shown on the diagonal, following our convention.

CHAOS IN DISCRETE DYNAMICAL SYSTEMS

The fixed point Q is repelling, while the fixed point P, although initially attracting when created at $b = 0.25$, underwent a period-doubling (flip) bifurcation when b increased past approximately 0.7495, as described in Chapter 2. So at the current value, 1.0, P is a repellor. This flip is the first in the Myrberg sequence, well studied in the works of Feigenbaum. In the box-within-a-box language of Mira, this flip is the opening of the first box of the second kind, an small interval of the b space, within a box of the first kind, a larger interval. Each box of the first kind includes a related box of the second kind.

FIGURE 7-1.

$b = 1.0$

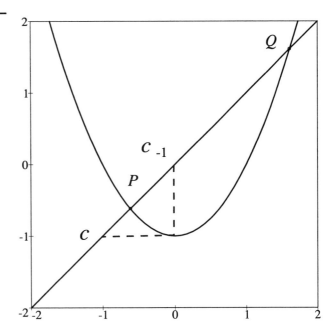

Stage 2: $b = 1.75$

A k-periodic point of the map f is a fixed point of the map f^k (f iterated k times). For example, to find a 3-cycle of f, we look for three related intersections of f^3 with the diagonal. In Figure 7-2, we see a fold bifurcation of f^3 occurring. The fixed points, P, Q, of f are still fixed points of f^3, so they appear here as intersections, as marked. But there are three new contacts as well, labelled $\alpha_1 = \beta_1$, $\alpha_2 = \beta_2$, and $\alpha_3 = \beta_3$. These comprise a 3-cycle of f, as well as fixed points of f^3. A box of the first kind opens here.

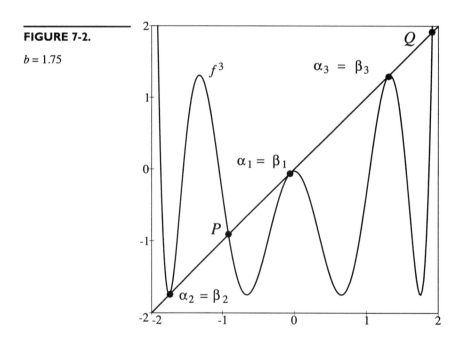

FIGURE 7-2.

$b = 1.75$

Stage 3: *b* = 1.76

Just after the fold bifurcation, there are two 3-cycles, one $\{\alpha_1, \alpha_2, \alpha_3\}$ attracting, the other $\{\beta_1, \beta_2, \beta_3\}$ repelling, as described in Chapter 2 and shown in Figure 7-3.

FIGURE 7-3.

b = 1.76

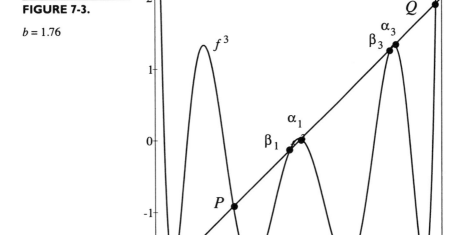

Stage 4: $b = 1.771$

At this value of b, an attracting 6-cycle has appeared recently, when the formerly attracting 3-cycle of the preceding stage flipped and became a repelling 3-cycle, opening the related box of the second kind. Figure 7-4 shows the attractive 6-cycle, both on the graph of f and on the diagonal.

Figure 7-5 shows a small rectangular neighborhood of the repelling 3-periodic point α_3, with two 6-periodic points nearby, both attracting. They are all shown on the crossing of the diagonal with the graph of f^6, as they are fixed points of this function.

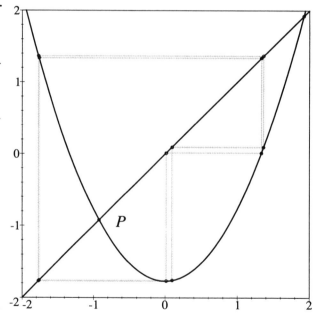

FIGURE 7-5.

$b = 1.771$

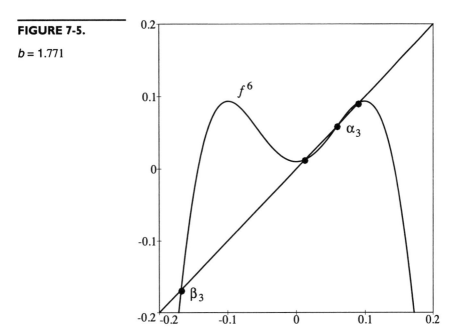

Stage 5: $b = 1.781$

Figure 7-6 shows that as b increases a bit more, the orbit shown seems to be chaotic in six disjoint intervals, rather than cyclic. Figure 7-7 is an enlargement of the pair of chaotic intervals in the center of Figure 7-6, which nearly abut the 3-periodic repellor, α_3. This enlargement shows evidence that a attractor fills the six intervals, and thus is chaotic (not cyclic).

FIGURE 7-6.

$b = 1.781$

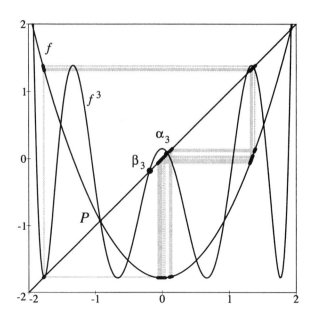

FIGURE 7-7.

$b = 1.781$

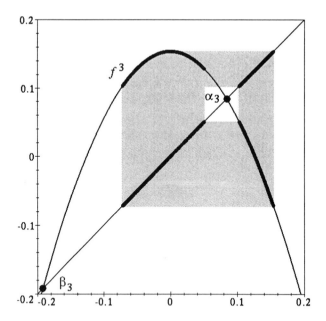

Stage 6: $b = 1.7822$

This is the moment of contact bifurcation. A critical point has moved toward, and now makes contact with, the point α_3 of the repelling 3-cycle $\{\alpha_1, \alpha_2, \alpha_3\}$. In this stage we see the closure of the box of the second kind which opened with the flip bifurcation of the α cycle, discussed in Stage 4.

Figure 7-8 shows a blowup of the same piece of f^3 seen in Figure 7-7. Here we see a trajectory of the basic critical point c_{-1} through three iterations of f^3 (that is, c_{-1}, c_2, c_5, c_8) until c_8 falls on α_3. Also, four preimages of c_{-1} are shown, indicating that a backward sequence of critical points, $c_{-4}, c_{-7}, c_{-10}, \ldots$, approaches α_3 asymptotically. The trajectory is shown in the Koenig-Lemeray style explained in Chapter 2.

With this contact there appear for the first time orbits homoclinic to α_3. These are points with orbits which are initially repelled by, but later jump back onto, α_3. A point like α_3 that has a homoclinic trajectory is called a *snap-back repellor* (SBR). In this figure we show three points c_{-1}, c_2, and c_5, which are homoclinic to α_3. They lie on opposite sides of the SBR α_3.

Before this bifurcation, the attractor occupied six chaotic intervals, permuted cyclically by the map f (and this is therefore called a 6-cyclic chaotic attractor). After this bifurcation, as we will see in the next figures, the attractor occupies only three (larger) chaotic intervals, permuted cyclically by the map f (a 3-cyclical chaotic attractor). Thus, we have at this instant an example of a CCB of the second kind, as described at the beginning of this chapter. The six intervals have joined pairwise into three larger intervals.

We may also see, in Figure 7-8, an example of an absorbing interval. This an interval bounded by critical points, which is mapped into itself and which is absorbing in the sense that all points sufficiently near will jump into the interval in a finite number of iterations. In this figure, the interval $[c_5, c_2]$ is absorbing. It contains a chaotic attractor of f^3. Also, it is invariant under the map f^3, that is, it is mapped exactly onto itself. On the other hand, under the map f, the three intervals — $[c_5, c_2]$ and its two images under the map, f — comprise an absorbing set which contains the 3-cyclic chaotic attractor. We also call these three component intervals of the absorbing set, collectively, *absorbing intervals*.

Note: The intervals are not individually mapped into themselves by f, but only by f^3. Their union is mapped into itself by f. Note also that before this homoclinic bifurcation, which is a CCB of the second kind, critical points define the boundary of 2·3-cyclic[1] absorbing intervals, which include the 2·3-cyclic attractor. The points α_i of the repelling 2·3-cycle, together with their preimages of all ranks, define the boundary of the *basin of attraction* of the 2·3-cyclic absorbing intervals, and each α_i separates two *immediate basins*. These concepts have been introduced in Chapter 2. When this CCB of the repelling 3-cycle occurs, the map has 6-cyclic chaotic intervals, not distinct, as well as 3-cyclic chaotic intervals. That is, the map f^6 has six invariant chaotic intervals, not disjoint, and the map f^3 has three disjoint invariant chaotic intervals.

The situation is similar for any CB of the second kind, in which a $2k$-cyclic chaotic attractor is transformed into a k-cyclic chaotic attractor: the *closure* of a box of the second kind is characterized by the appearance (for the first time) of homoclinic orbits of the k-cycle, α, on both sides of the points α_i of the cycle, and the $2k$ absorbing intervals merge into k absorbing intervals. This occurs without an abrupt increase in the size of the absorbing set, such as we will see in the next example.

So far we have introduced most of the main ideas pertaining to CCBs of the second kind, which closes a box of the second kind. Next we look at a CCB of the first kind, which closes a box of the first kind opened in Stage 2 above.

---

1. This means 6-cyclic, and recalls that $6 = 2 \cdot 3$.

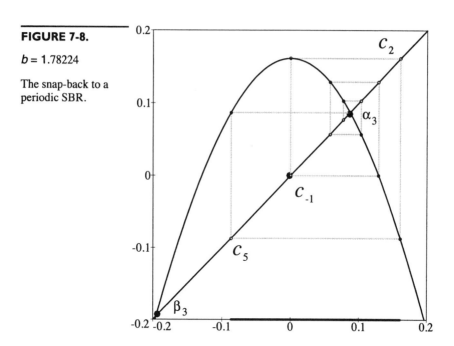

FIGURE 7-8.

$b = 1.78224$

The snap-back to a periodic SBR.

Stage 7: $b = 1.79032$

As in the preceding CCB of the second kind, this CCB of the first kind is characterized by the merging of critical points into a k-cycle, which was born repelling at the time of the opening of the box of the first kind. In this case, the affected cycle is the 3-cyclic attractor, β, born in Stage 2 above.

Figure 7-9 shows the 3-cyclic chaotic attractor, in the context of the graph of f^3, with the points α_3 and β_3 on the diagonal, at the moment of a CB of the first kind. Figure 7-10 is an enlargement of the region containing β_3 and α_3 in Figure 7-9. Here we can see that a critical point, c_5, has merged onto β_3, and is the terminus of a homoclinic orbit of critical points.

FIGURE 7-9.

$b = 1.79032$

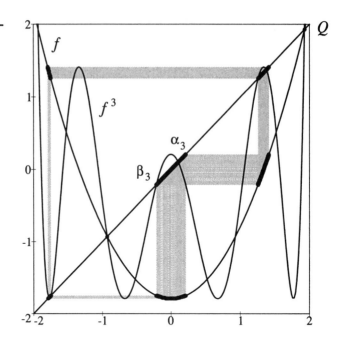

FIGURE 7-10.

$b = 1.79032$

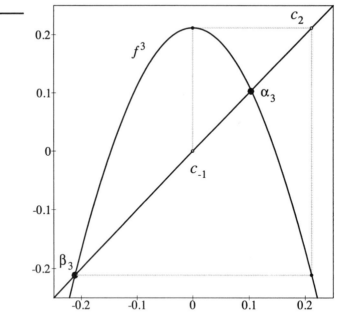

Stage 8: *b* = 1.7905

Figure 7-11 shows that after a CCB of the first kind, the chaotic attractor has exploded into a single interval. Figure 7-12 is a view of the chaotic attractor alone, in the same frame. This is significantly different from the case of the CCB of the second kind described above. To explain this difference, consider again the fold bifurcation of Stage 2, in which the α and β 3-cycles were created, and in which the box of the first kind opened. Looking back at Figure 7-2, we can see that, at the moment of fold bifurcation, the points of the newborn cycle are attracting from one side (slope steeper than 1) while repelling to the other (slope less steep than 1). (This may be verified by drawing some cobwebs.) Still at the moment of this bifurcation, the new cycle has homoclinic orbits on the repelling side.

FIGURE 7-11.

b = 1.7905

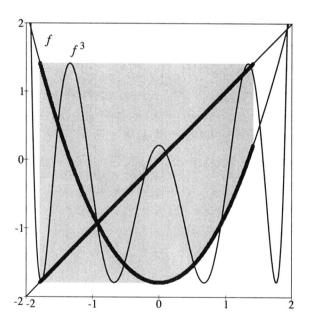

FIGURE 7-12.

$b = 1.7905$

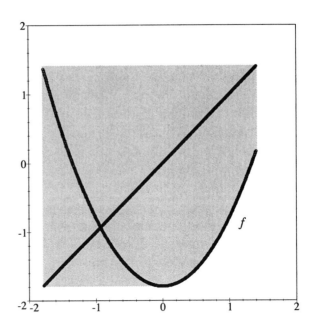

Stage 2: (again): $b = 1.75$

Figure 7-13 shows an enlarged view of the region near $\beta_3 = \alpha_3$ at the moment that point is created in the fold bifurcation. This point is repelling to the left, and two homoclinic orbits are shown. There are infinitely many such orbits. This point is homoclinic on one side only. Homoclinic orbits are also shown in Figure 7-14, *after* the fold bifurcation.

FIGURE 7-13.

$b = 1.75$

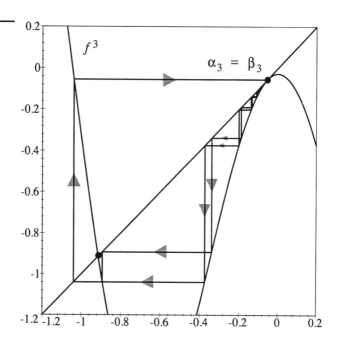

Stage 3: (again): $b = 1.76$

Looking back at Stage 3 in Figure 7-14, we see that β_3 still has infinitely many homoclinic orbits on the left, and none on the right. This means that the basin of attraction of the attractive 3-cycle, α, or of the cyclic absorbing intervals existing before the CCB of the first kind, is made up of the immediate basins together with all of their preimages, which have a chaotic, or fractal, structure. However, once a point of a trajectory enters the immediate basin bounded by β_3 and its preimage, $(\beta_3)_{-1}$, its images will enter the absorbing interval and never escape, as may be seen in Figure 7-8.

Summarizing the CCB of the first kind of Stage 6, the chaotic interval has a contact with β_3, and homoclinic points appear also on the other side of that point, as shown in Figure 7-10.

FIGURE 7-14.

$b = 1.76$

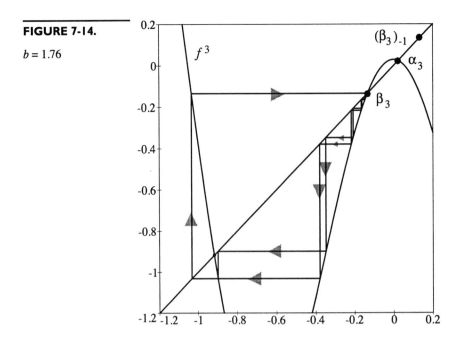

Figure 7-15 shows that after a CCB of the first kind (this does not occur in a CCB of the second kind) the generic trajectory of a point of the former immediate basin, $]\beta_3, (\beta_3)_{-1}[$, will escape from that open interval, covering the whole absorbing interval, $]c, c_1[$. In particular, soon after this CCB, the generic trajectory spends more time in the former chaotic intervals, but some rare moments outside them.

FIGURE 7-15.

$b = 1.79035$

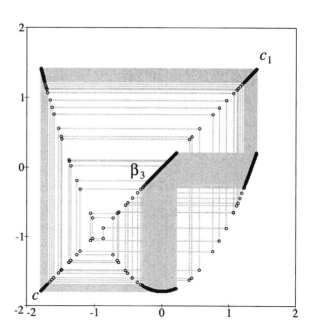

7.2   BIFURCATIONS IN TWO DIMENSIONS

We now introduce chaotic contact bifurcations (CCBs), for maps of the plane. The introduction to CCBs in 1D of 7.1 will have provided a preparation for this material. The fine structure of CCBs may be easier to visualize in 2D than in 1D.

Context

Our context is a one-parameter family of maps of the plane into itself. Each map of the family is the generator of a discrete dynamical system based on the iterations of the map. As the parameter is changed, the dynamics (attractors, basins, and so on) of the map change. Special changes are known as bifurcations.

In this chapter we consider maps having a chaotic attractor. CCBs involve qualitative changes, or even destabilization, of the shape of a chaotic attractor. This has been a subject on the frontier of discrete dynamical systems theory since GM2 in 1978.

What is a CB?

A *contact bifurcation* involves a contact between the boundary of a chaotic attractor and the boundary of its basin of attraction. Both of the boundaries involved in the contact may be fractal. The underlying causes of fractal (rough) basin boundaries comprise a main theme of dynamical systems theory, and several mechanisms may be seen in the exemplary bifurcation sequences of this book.

In one of these exemplary sequences, we will explain, step-by-step, a bifurcation sequence taken from GARF. In this sequence, the CCBs correspond to homoclinic bifurcations of repelling cycles (nodes, foci, or saddles) of the map. In preparation for the detailed explanations of those examples, we will introduce here some of the basic concepts, and some of the technical jargon, of CCB theory.

Basic concepts and notations

This is an informal introduction; precise definitions may be found in Appendices 2 and 3. A subset of the plane is *invariant* under the map if the subset is mapped exactly onto itself. A *chaotic area* is an invariant subset (larger than a finite set) that exhibits *chaotic dynamics*, that is, a typical trajectory fills the area densely. A subset of the plane, A, is *attracting* if it has a neighborhood (an open set, U, containing A) every point of which tends asymptotically to A, or arrives there in a finite number of iterations. In this case, the basin or *basin of attraction* of A, denoted $D(A)$ or D, is the set of all points which eventually enter A; this basin may be found by taking the union of all preimages of the neighborhood U. The *immediate basin* of A, denoted $D_0(A)$ or D_0, is the largest connected part of D containing A. A *chaotic attractor* is a chaotic area which is attracting.

Consider now a map T with a chaotic attractor, d, its basin D, and the boundary of this basin, $F = bd(D)$, also called the *frontier* of d.

The attractor is called a *k-cyclic chaotic attractor*, where k is a positive integer, when d has k connected components which are permuted cyclically by the map T. Let T^k denote the map T applied k times in succession. Then each component of the attractor d of T is individually an attractor of T^k. In this case, the basin D is also k-cyclic. That is, it has k connected components, each being an image of the immediate basin, which are permuted cyclically by T. Each basin component, and each attractor component, are invariant under the map T^k. This applies to F as well.

What is a CCB?

A *chaotic contact bifurcation* occurs in this context when, as the parameter of the family is varied, the chaotic attractor d moves toward its basin boundary, the frontier F, and eventually makes contact when the attractor boundary, $bd(d)$, touches F. This frontier may be either a fractal, or smooth but the limit of an infinity of folded loops compressed by the action of the map. Also, the frontier contains repelling cycles, either saddles or repellors of nodal or focal type.

If d is k-cyclic, $k > 1$, we use T^k instead of T to visualize this event more easily. Let d_0 be one of the components of d. Then the attractor of T^k, d_0, will drift toward its basin boundary $F_0 = bd(d_0)$.

Classification of CCB types

A point P of F is an *isolated point* of F if it has a neighborhood, every point of which is not in F.

A *snap-back repellor* (SBR) is a repelling node or focus, P, which has a *homoclinic point*, that is, a point Q which has preimages approaching asymptotically to P, and also has an image which is P. That is, the homoclinic point Q comes from P in the infinite past, and arrives at P in the finite future. Hence, the orbit of Q is homoclinic, or "same-tending", in the sense of tending to the same point, P, in the past and in the future.

We classify CCBs as type I or type II. Those of type II are further divided into three kinds: first kind, second kind, and final.

In a *CCB of type I*, the contact of $bd(d)$ and F is first made at a fixed point, P, which is an isolated point of F until the moment of contact. Usually, this contact point becomes an SBR, in which case the CCB is called a *homoclinic bifurcation*. At the instant of a CCB of type I, the boundary touches an SBR, and the homoclinic orbits make their first appearance at P at the moment of contact. In this case, the point P is no longer within F at the moment of contact.

In a *CCB of type II*, the contact of $bd(d)$ and F occurs at points which were not isolated points of F before the moment of contact. These are also usually homoclinic bifurcations.

The three kinds of CCB of type II

A CCB of type II is *of the first kind* if it causes a sudden change in the shape of a chaotic attractor, such as a sudden change in size.

A CCB of type II is *of the second kind* if it causes a qualitative change in the shape of a chaotic attractor, such as the joining of a finite number of chaotic sets into a smaller number of chaotic sets, which continue to attract after the bifurcation.

A CCB of type II is a *final bifurcation* if it destroys the attractor, that is, it changes from a chaotic attractor to a chaotic repellor.

If F_0 is a limit set of components of basin boundaries of other bounded attractors (that is, consists of limit points of sequences of points belonging to basin boundaries) then a CCB of type II of the first kind may occur. If the contact points belong to the immediate basin of another bounded attractor, then a CCB of type II of the second kind may occur.

The simplest case in which a CCB of type II is a final bifurcation occurs when the contact points are limit points of the basin of infinity (the set $D(\infty)$ of points having unbounded trajectories). However, other similar interactions may give rise to this kind of bifurcation, in which $D(\infty)$ is not involved, but another basin, D', plays its role.

Next, we give examples, with abundant graphics, of CCBs of type II of the first and second kinds. Other examples may be found in G1, GARF, and FMG. Several examples of final bifurcations have already been described – see the bifurcation at the last stage described in each of the preceding three chapters.

7.3   EXEMPLARY BIFURCATION SEQUENCE

We have introduced the basic concepts and notations of CCBs in 1D in Section 7.1, and in 2D in Section 7.2. Here, we continue the 2D discussion with an exemplary bifurcation sequence, showing CCBs of type II of the first and second kinds.

We use a quadratic family of maps of the unit square, [0, 1] x [0, 1], called the *double logistic family*; this is EQ2 in 1.5, for more details, see GARF and FMG. As the bifurcation parameter increases from 0.6 to 1.0, this family exhibits features of a family of one-dimensional quadratic maps — the logistic family, equivalent to the Myrberg family of Section 7.1 — on the *diagonal*, the set Δ of points of the form (x, y) with $x = y$. This set is mapped into itself by every map of the double logistic family. Another special feature of these maps is a symmetry with respect to reflection through the diagonal. Also, each map has four fixed points: the points $(0, 0)$ and $(0.75, 0.75)$ on the diagonal, and two other points, P_1^* and P_2^*, which are mirror reflections through the diagonal.

Skipping over some simple bifurcations for low values of the bifurcation parameter b, we come to a type II CCB of the second kind between 0.64218 and 0.64219.

Stage 1: $b = 0.641$

Just before the bifurcation, the double logistic map has a 14-cyclic chaotic attractor (the dark curves in Figure 7-16), which belongs to an annular absorbing area (shown in white in Figure 7-16).

In the enlargement of Figure 7-17, we see a 7-periodic point of saddle type, V, between two nearby chaotic areas, shown with its local stable curve or inset, W^s, and its unstable curve or outset, W^u. To fix these features, we should think in terms of the map T^{14}. The inset of V is the frontier between the basins of the two chaotic attractors. Note that the two rays of the outset of V are attracted to the two chaotic attractors, and approximate their shapes.

FIGURE 7-16.

The basin of infinity in gray; the attractor is black.

$b = 0.641$

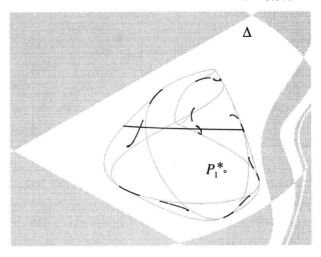

FIGURE 7-17.

$b = 0.641$

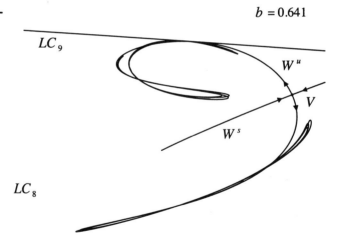

Stage 2: $b = 0.64218$

Just an instant before the bifurcation, there is almost contact between the two chaotic areas and the frontier, the inset of V, as shown in Figure 7-18. Figure 7-19 is an enlargement of the small square in Figure 7-18. It shows how closely the chaotic attractors have approached to the frontier.

FIGURE 7-18. $b = 0.64218$

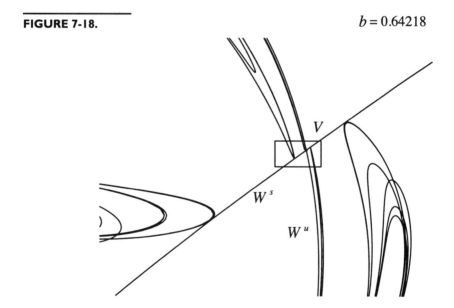

FIGURE 7-19.

$b = 0.64218$

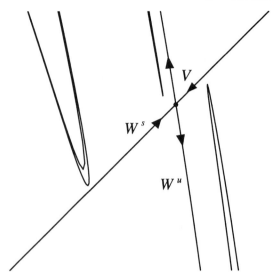

Stage 3: $b = 0.64219$

Just an instant after the bifurcation, the chaotic attractors have passed through the inset of V, as shown in Figure 7-20. Because the outsets of V approximate these two attractors, which cross transversally through the inset of V, we may conclude that V is a transversally homoclinic saddle cycle: its outset has infinitely many transversal intersections with its inset. In fact, V has experienced a transition from nonhomoclinic to homoclinic state, exactly at our contact bifurcation.

FIGURE 7-20.

$b = 0.64219$

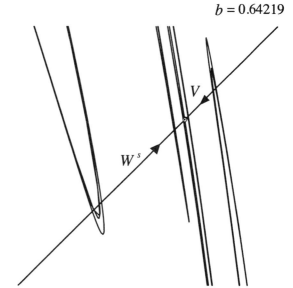

Stage 4: $b = 0.643$

Later on, as shown in Figure 7-21, we see that the 14-cyclic chaotic attractor has become a 7-cyclic chaotic attractor, through the merging of pieces in pairs. This is characteristic of the CCB of type II of the second kind.

FIGURE 7-21.

$b = 0.643$

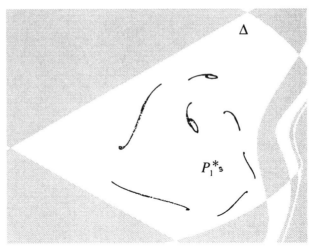

Stage 5: $b = 0.6439200$

This stage represents the situation just an instant before a type II CCB of the first kind. In this event, the 7-cyclic chaotic attractor will explode into a larger, annular, connected (that is, 1-cyclic) chaotic attractor, which will be seen clearly later at stage 7. Such an explosion is characteristic of a type II CCB of the first kind.

In Figure 7-22, we see the complex dynamics just before this explosion. The attractor is shown in black, and the seven components of the basin of the 7-cyclic chaotic attractor are shown in 7 shades of grey, in Figure 7-22. (Please note: The seven components of one basin of the map T are seven distinct basins of seven distinct attractors of the map T^7.)

Figure 7-23 is an enlargement of the rectangle indicated in Figure 7-22, showing the close approach of a chaotic attractor of T^7 to its basin boundary. There is a 21-cycle of T of saddle type, that is, a 3-cycle saddle of T^7, belonging to the boundary of the immediate basin. This basin is shown in the lightest shade of grey.

FIGURE 7-22.

To each part of the attractor (black) is associated a corresponding piece of the basin (shades of gray.) One piece of the fundamental critical curve and two pieces of the basic critical curve are shown.

$b = 0.64392$

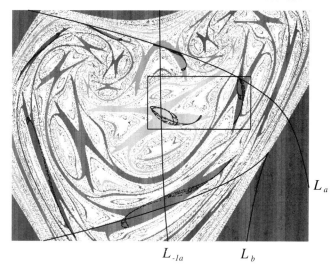

L_a

L_{-1a} L_b

FIGURE 7-23.

Enlargement of the small rectangle above.

$b = 0.64392$

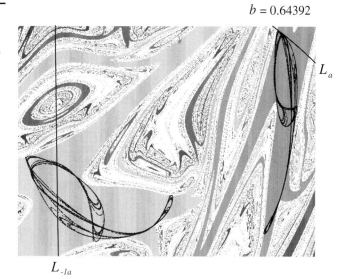

L_a

L_{-1a}

Stage 6: *b* = 0.6439248

Just after this contact bifurcation, the attractor has firmly pierced its former frontier. Figure 7-24 is an enlargement near one of the saddle points of the 21-cycle. This saddle lies on the former basin boundary, which is the inset of the saddle, and the attractor passes through this former frontier. Figure 7-25 is the same view, without the basins. Here we clearly see that the saddle has become homoclinic, as the unstable set crosses the inset of the saddle, and the chaotic attractor crosses the former frontier.

FIGURE 7-24.

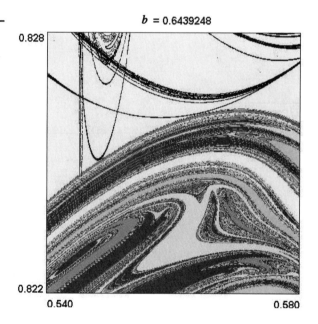

b = 0.6439248

0.828

0.822

0.540   0.580

FIGURE 7-25.

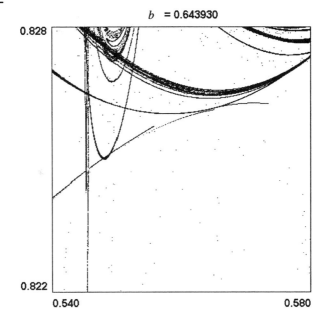

b = 0.643930

Stage 7: b = 0.644

Still later, we see the outcome of this CCB of type II of the first kind. Figure 7-26 shows the large, connected, annular chaotic attractor. It overlays the area formerly occupied by the seven pieces of the 7-cyclic chaotic attractor.

This large attractor persists until b reaches 0.702. Of course, it has a mirror image on the other side of the diagonal, due to symmetry, so there are actually two large chaotic attractors. The two corresponding immediate basins are separated by a segment of the diagonal, which segment consists of the inset of a 2-periodic cycle, $\{Q_1, Q_2\}$, which exists on the diagonal along with the fixed points of the map.

FIGURE 7-26.

$b = 0.644$

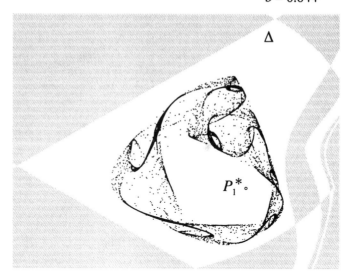

Δ

$P_1^* \,\circ$

Stage 8: *b* = 0.7020

A change suddenly occurs at this stage. The large annular chaotic attractors have expanded, and now make contact simultaneously with the diagonal. Figure 7-27 shows the points Q_1, Q_2, on the diagonal, Δ, and the contact just established with one of the large attractors. This is a CCB of type II of the second kind. The two attractors are becoming one, which will be symmetric, of course. At the bifurcation, the periodic points, Q_1, Q_2, are critical; that is, they belong to critical curves.

Stage 9: *b* = 0.7025

Just after the CCB, the outset of the 2-cycle, $\{Q_1, Q_2\}$ crosses the inset of the 2-cycle, which is in Δ, as shown in Figure 7-28. Thus, this 2-periodic saddle has become homoclinic during the CCB.

FIGURE 7-27.

$b = 0.702$

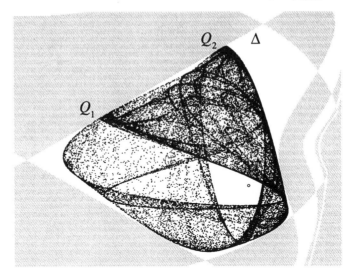

$b = 0.7025$

FIGURE 7-28.

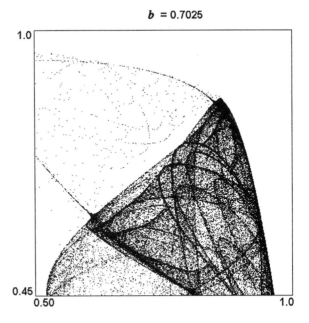

Stage 10: $b = 0.7030$

Figure 7-29 gives a global view of the new one-piece chaotic attractor. Note the three holes, labelled $H(S*)$, $H(P_1*)$, and $H(P_2*)$, which are bounded by critical curves. These surround the fixed repelling node on the diagonal, $S*$, and the symmetric repelling foci, P_1* and P_2*. These three holes will disappear at CCBs of type I, which are the first homoclinic bifurcations.

Stage 11: $b = 0.714$

At this CCB, the hole $H(S*)$ disappears. Figure 7-30 shows the critical curves at this CCB. Figure 7-31 shows the attractor at the same moment.

FIGURE 7-29.

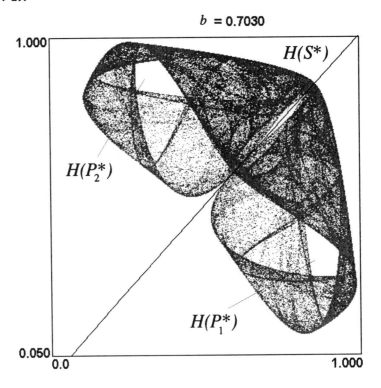

$b = 0.7030$

CHAOS IN DISCRETE DYNAMICAL SYSTEMS

Stage 12: $b = 0.7375$

At this CCB, the holes $H(P_1{}^*)$ and $H(P_2{}^*)$ disappear. Figure 7-32 shows the critical curves at this CCB, Figure 7-33 shows the attractor, now simply connected, at the same moment.

FIGURE 7-30. $b = 0.714$

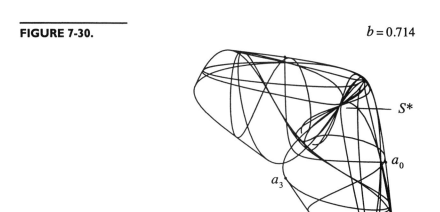

FIGURE 7-31.   b = 0.714

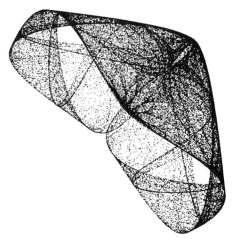

As b continues to increase, several other CCBs of type I or type II may be observed, in which the sudden change of shape of a chaotic attractor is similar to the changes already seen in this sequence. Watch for them in the movie of the full bifurcation sequence provided on the CD-ROM which accompanies this book. We especially recommend $b = 0.88$, $b = 0.88498$, and $b = 0.88499$.

FIGURE 7-32.

$b = 0.7375$

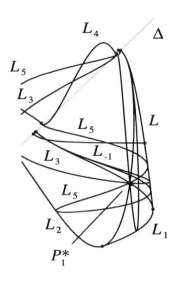

FIGURE 7-33.

$b = 0.7375$

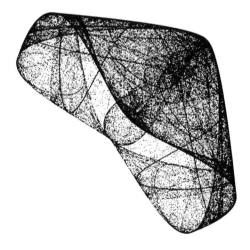

CHAOS IN DISCRETE DYNAMICAL SYSTEMS

CHAPTER 8

CONCLUSION

This completes our tour of attractors and bifurcations in the context of discrete dynamical systems in two dimensions. Most of the basic concepts — attractors, basins, critical sets, bifurcations, and so on — may be understood in the 1D context, as we have indicated here and there; but perhaps they are clearer in 2D. Also, the 2D versions may admit a more straightforward generalization to 3D and higher dimensions.

We have presented this new branch of mathematics as an experimental subject; for a young subject, that is most appropriate. As time goes on, chaos theory in the context of discrete dynamical systems may gain a more formal framework. Meanwhile, we encourage you to use our software, or your own, to explore the new frontier!

PART 3

APPENDICES

The appendices provide some additional details that were omitted from the text in order to keep the main ideas as clear and simple as possible. Especially, we have collected the definitions expressed in mathematical notation (Appendices 1-3) and some additional historical information (Appendices 5-6).

APPENDIX I

NOTATIONS

The following lists define some symbols that we use in the Appendices.

AI.I FORMAL LOGIC

$\forall$: for all
$\exists$: there exists
$\ni$: such that

AI.2 SET THEORY

$\{x \mid P(x)\}$: the set of x such that $P(x)$ is true
N: the natural numbers, $\{0, 1, 2, \ldots\}$
$\mathbf{Z}$: the integers, $\{\ldots -2, -1, 0, 1, 2, \ldots\}$
$A \subset M$: A is a subset of M
$A \cup B$: the union of A and B
$A \cap B$: the intersection of A and B
$M \backslash A$: for $A \subset M$, the complement of A in M
$f : A \to B$: f is a function from A to B
$f / C : C \to B$: the restriction of f to a subset of C of A

AI.3 POINT SET TOPOLOGY

$\overline{A}$: for $A \subset M$, the topological closure of A (in M)
$A°$: the interior of A
∂A: for $A \subset M$, the topological boundary of A,

$$\partial A = \overline{A} \cap \overline{M \backslash A}$$

TOPOLOGICAL DYNAMICS

We assume a familiarity with the fundamentals of point-set topology and vector calculus, and refer to manifolds sometimes, but global analysis is not required. Our goal in this section is to build a bridge from the elementary level of the earlier chapters to the textbooks of pure mathematics.

Let M be a differentiable manifold. (We may simply imagine here an open set of Euclidean space, or a sphere or torus, for example.) By an *endomorphism* of M we will always mean a smooth function from M to itself. In fact, our theory makes use of continuous, piecewise-differentiable functions, but in this appendix we consider only the smooth case, for simplicity.

A2.1 TRAJECTORIES AND ORBITS

A *map* is a continuous function of topological spaces. A function on a differentiable manifold M is a *diffeomorphism* iff it is smooth, invertible, and its inverse, $f^{-1}:M \to M$, is smooth. A diffeomorphism generates a *cascade* (that is, an action of the integers, $\mathbf{Z}$), as follows. Let f^0 denote the identity map, and $f^1 = f$. For an integer k, let $f^k(x) = f(f^{k-1}(x))$. For any $k \in \mathbf{Z}$, let $x_k = f^k(x)$. Then the map $k \to f^k$ is a cascade, that is, a group homomorphism from $\mathbf{Z}$ to the diffeomorphism group of M. The *trajectory* of a point x is the doubly infinite sequence, (x_k), with $k \in \mathbf{Z}$. The *orbit* of x is the image (that is, the underlying set) of its trajectory, denoted by $o^*(x)$. The trajectory is an ordered sequence, the orbit is a set.

In the case of a semi-cascade (that is, an action of the natural numbers, N) generated by a noninvertible map, the trajectory of a point is a singly infinite sequence, (x_n) with $n \in N$. Again, its orbit is the image of its trajectory, denoted by $o^*(x)$. Similarly, we may speak of the trajectory and orbit of a subset of M.

A2.2 INVERSE IMAGES

For a noninvertible map $f:M \to M$, an image $f(x)$ is a point of M, but an inverse image $f^{-1}(y) = \{x \mid f(x) = y\}$ is a subset of M. As before, we denote the iterated image points by $x_n = f^n(x)$, for n a natural number. But now, the iterated *inverse* images are subsets of M, denoted by $f^{-k}(y) = \{x \mid f^k(x) = y\}$. Such a point, x, is called a *preimage of y of rank k*.

In this theory we will be particularly concerned with those endomorphisms that behave like generic polynomial maps. We define these special maps as follows.

Definition. A function $f:M \to M$ is *finitely folded* if M is a finite union of closures of open subsets, $M = \overline{M}_1 \cup \dots \cup \overline{M}_p$, such that the restricted maps, $f_j = f/M_j$, are homeomorphisms.

In the case of a nice, noninvertible map, the notations f_j^{-1}, $j = 1,\dots, p$, denote the inverses of the restricted maps, which are homeomorphisms, and these are regarded as partial inverses of the noninvertible map f.

A2.3 FIXED POINTS

We fix now an endomorphism, $f:M \to M$, and its semi-cascade. In this context, a *fixed point* is a point x of M such that $f(x) = x$. As in the case of a diffeomorphism, the stability of a fixed point is generically determined by its *characteristic multipliers*, that is, the eigenvalues of the *derivative* of the map, which is the linear endomorphism, $T_x f:T_x M \to T_x M$, which best approximates the map in a neighborhood of the point x. A fixed point x of f is *weakly stable* iff for every neighborhood W of x there is a neighborhood U of x, with $U \subset W$, such that the orbit of U is contained in W. (This is the *stable fixed point* of G1, p. 6. But here, stability is defined in 2.6 below.) A fixed point which is not weakly stable is called *unstable*.

A2.4 PERIODIC TRAJECTORIES

In the case of a cascade, one of the important qualitative features is a *periodic trajectory*, that is, a trajectory whose orbit contains only a finite number of points. If k is the number of points in its orbit, the trajectory (and each of its points) is said to have *prime period k*. Periodic trajectories are also an important qualitative feature of semi-cascades, and their classification introduced by Poincaré in the context of cascades (nodes, foci, saddles, etc.) applies equally to semi-cascades. The *characteristic multipliers* of a k-periodic trajectory, orbit, or point, are the characteristic multipliers of the corresponding fixed points of the endomorphism f^k, the k-fold iteration of f.

A2.5 LIMIT POINTS

The notion of an ω-limit set from cascade theory applies equally well to a semi-cascade. For a point $x \in M$, the ω-*limit of x* is the set of limit points of the trajectory of x as a sequence, or equivalently,

$$\omega(x) = \bigcap_{n \in N} \overline{o(x_n)},$$

the intersection of the closures of the orbits of the images of x under the endomorphism.

A2.6 STABLE SETS, ATTRACTORS, AND BASINS

Let A be any subset of M. In the context of a fixed endomorphism f, A is *trapping* iff $f[A] \subset A$, and *invariant* iff $f[A] = A$. Consider now a set A which is closed and invariant. Then A is *stable* iff every neighborhood W of A contains a neighborhood U of A

which is trapping. The *stable set* of A, $W^S(A)$, is the set of all x of M such that the ω-limit set of x belongs to A. A closed, invariant set $A \subset M$ is *attractive*, or *attracting*, iff it is stable and it has a trapping neighborhood in the interior of its stable set. An *attractor* is an attractive set which is irreducible in the weakest sense: it has a dense orbit. If A is attractive, the stable set of A is called the *basin* of A.

Note: If U is a neighborhood of A as in the definition of attractive set, then the basin of A is the union of the preimages of U.

A2.7 UNSTABLE SETS AND REPELLORS

Let A again be a closed, invariant set. The *unstable set* of A, $W^u(A)$, is the set of all points having a sequence of preimages converging to a subset of A. The set A is *repelling* iff it is not attracting.

Note: Here we diverge from common usage in cascade theory, in which repelling means the opposite of attracting. Instead, we use the following notion. The closed, invariant set, A, is *expanding* iff every neighborhood W of A contains a neighborhood U of A which is contained in its own image, and such that every point of U has a sequence of preimages converging to a point in A. For example, a fixed point which is a saddle is repelling, and one which is a repelling node or focus is expanding.

A2.8 CHAOTIC ATTRACTORS

Let A be a closed, invariant set. Then the action of f on A is *sensitive*, or has *sensitive dependence on initial conditions* iff there exists a $\beta > 0$ such that, for any point x of A, and any neighborhood U of x, there is a point $y \in A \cap U$ and a positive integer n such that $d(x_n, y_n) > \beta$. The action of f on A is *topologically transitive* iff for any two open sets U, V, of A, there is a positive integer n such that $f^n[U] \cap V \neq \emptyset$. Further, A is called a *chaotic set* under the action of f iff it is sensitive, topologically transitive, and the set of all periodic points is dense in A. Finally, a *chaotic attractor* is an attractor that is a chaotic set.

CRITICAL CURVES

The method of critical curves, discovered by Mira in 1964 and presented originally in GM1, has proved vital to the geometric theory of semi-cascades, as it has evolved up to now. In fact, one of the chief features of a noninvertible endomorphism, which distinguishes it from a diffeomorphism, is the existence of critical curves. In this section, we introduce Mira's nomenclature for the critical curves, following Appendix A of (Gardini, 1992b). At this point, we restrict the manifold, M, to the plane.

A3.1 THE ZONES

We now decompose the domain of the semi-cascade — the plane — into sets of points all having the same number of pre-images of rank 1, or *layers*, under the folding action of the endomorphism. For a natural number n, let L_n denote the set of all points for which the cardinality of the inverse image is n. (Do not confuse with the critical curves.) For $n = 0$, this set is the complement in the plane of the image of the map. Let L_∞ denote the set of points of the plane for which the inverse image is not a finite set. Then M is decomposed into a finite or countably infinite union of disjoint sets, called *layer sets*. In the usual examples we will consider, the endomorphism is nice, the layer set decomposition is finite, and the set L_∞ is empty. Therefore, we make the following restriction of our endomorphism, f.

Finite-zone assumption: The endomorphism is nice, and the inverse image of every point is a finite set; that is, $L_\infty = \varnothing$.

For example, a nice quadratic map may have layer sets for $n = 0, 1,$ and 2, which are open sets only for $n = 0$ and 2. The layer set for $n = 1$, in such a case , is a curve dividing the plane into the two opens sets with 0 and 2 layers, respectively. This is the fold curve in the classic language of singularity theory. A map with these properties is said to be *of type* $Z_0 - Z_2$, or *of type 0 – 2*. In a context in which there is more than one map, we write $Z_n(f)$ for the zones of the map f.

A3.2 CRITICAL POINTS VIA CALCULUS

The critical curves of an endomorphism are fundamental to our theory, and here we will give two different definitions of them, via calculus and via topology. The two systems of definitions are not equivalent in general, but they do agree for generic smooth mappings. (We do not prove this here.) We proceed now, in a sequence of steps, with the differential calculus version.

A (vector) point x of the domain of the map f is said to be a *regular point* of the map iff the derivative (linear map) $T_x f$ is injective. As in our context the derivative is a linear mapping from the plane to itself, this is equivalent to to saying that the derivative is a linear isomorphism. A point that is not regular is said to be a *critical point* of the map. Classically, the two generic possibilities for a critical point x (in the two-dimensional context) are that x is either:

- a *fold point* (rank of the derivative is 1, and the map is locally of type 0 – 2); or

- a *cusp point* (the rank of the derivative as a linear map is 1, and the map is locally of type 1 – 3 – 1).

A *critical value* of the map is an image of a basic critical point. Let $C_{-1}(f)$ denote the set of critical points of f, and $C_0(f)$ the set of critical values. Note that the inverse image of the set of basic critical values is much larger, in general, than the set of basic critical points; t is always the case, however, that $C_{-1}(f) \subset f^{-1}(C_0(f))$.

Generically, the critical points comprise a set of smooth curves without cusps, while the critical values comprise a set of smooth curves, possibly with cusps. The cusps are the images of the cusp points.

A3.3 CRITICAL POINTS VIA TOPOLOGY

In the literature on the critical curves of plane endomorphisms, the style of definition parallels the one adopted here, but we have modified the usual definitions slightly to allow a closer comparison with those of the previous section. Let f be a continuous plane endomorphism. Then a point x of the domain of the map is said to be a *regular point* of the map iff there is a neighborhood U of the point such that the restricted map, f/U, is injective. Otherwise the point is a *critical point*. It is immediate that, in the case of a smooth map, differential regularity implies continuous regularity, via the inverse function theorem. Consideration of a horizontal inflection point (even in one dimension) shows that the converse is false.

Continuing with our definitions, a *critical value* is an image of a critical point, and a *regular value* is any point which is not a basic critical value. A critical value is also said to have *coincident preimages*, since at least one of its preimages, a critical point, has no neighborhood on which the map is injective. Let L_{-1} denote the set of all critical points of f, and let L denote the set of all critical values. Note that by definition, $L = f[L_{-1}]$, while in general, $L_{-1} \subset f^{-1}[L]$.

It is now possible to relate these critical values to the zones Z_n of the map, under the finite zone assumption: the critical values belong to the boundaries of the zones. In the examples we usually use, the converse is also true, and the zone boundaries are identical to the set of all critical values.

Henceforth, we shall adhere to the definitions of this section for the critical points of a map. Note that the main difference is this: a point is regular in the sense of differential calculus if the map is locally a diffeomorphism, and regular in the sense of point-set topology if the map is locally a homeomorphism. A horizontal inflection point is critical in the first sense, and regular in the second sense.

A3.4 THE CRITICAL CURVES

We now define the all-important critical curves. We have L_{-1} and L as before, with $L = f[L_{-1}]$. We now define $L_1 = f[L]$, and in general, $L_{n+1} = f[L_n]$. These are the *critical curves* of the map f.

Generically, for smooth maps, the critical curves are finite unions of smooth curves, and the basic critical curve, L_{-1}, is the critical set in the sense of differential topology. See, for example, the classical work of Hassler Whitney (Whitney, 1955) on the singularities of plane endomorphisms.

A3.5 ABSORBING AREAS

An *absorbing area* is a closed and bounded subset A of the plane, such that

- A is *trapping*, that is, mapped into itself;

- A is *super-attracting*, that is, there is a neighborhood U of A such that for every point x of $U \setminus A$ there is a positive integer, n, such that trajectory of x enters the interior of A in n steps, that is, $x_n \in A^0$; and

- The boundary of A consists of a finite union of arcs of critical curves.

Note: Absorbing areas with more general boundaries could be defined, but we do not consider them in this book.

SYNONYMS

In order not to complicate the text with synonyms, we list them here. The term we commonly use is italicized. The phrase which follows it may be encountered in other works of chaos theory.

Bounded set: a set at finite distance.

Critical point: a critical point of rank 0, denoted LC_{-1}.

Critical value: critical point of rank 1, denoted LC.

Repellor: a point which is not an attractor.

Boundary: a frontier.

Diffeomorphisms: a smooth, invertible map with a smooth inverse.

Map: an endomorphism, homeomorphism, continuous function, or diffeomorphism, depending on the context.

HISTORY, PART I

This is a translation of the historical section of GM1, pp. 69 – 78, written in 1980, and is limited to the contributions prior to that date.

A5.1   EARLY HISTORY

Iterations have been used as a tool for a long time for solving various mathematical problems, such as:

- the definition of certain functions;

- the solution of algebraic equations by successive approximations;

- the definition of certain sequences.

Here are four examples.

Example 1. Hurwitz defined the Naperian logarithm of a number x_0 by means of the recurrence $x_{n+1} = \sqrt{x_n}$, of which the solution is

$$x_n = (x_0)^{2^{\frac{1}{n}}}$$

which tends to 1 as n increases. Thus,

$$ln(x_0) = \lim_{n \to \infty} 2^n(x_n - 1)$$

from which the properties of the logarithm may be recovered.

Example 2. To calculate the roots of the fixedpoint equation $x = f(x)$ one uses the recurrence

$$x_{n+1} = f(x_n).$$

If $f'(x)| < 1$, we find an interval containing a single root, which is the limit of the sequence $x_0, x_1, x_2, \ldots$

Example 3. Newton's method leads to a recurrence also. For if $y = F(x) = 0$ is the equation to be solved, we iterate the recurrence

$$x_{n+1} = x_n - F(x_n)/F'(x_n).$$

Example 4. The Fibonacci sequence may be defined by the recurrence relation

$$x_{n+2} - x_{n+1} - x_n = 0,$$

with the initial conditions, $x_0 = x_1 = 1$. In fact, the characteristic multipliers of the fixed point at zero are

$$S_1 = (1 + \sqrt{5})/2, \, S_2 = (1 - \sqrt{5})/2.$$

The first of these is the golden ratio of the ancients, and is the limit of the sequence of ratios x_{n+1}/x_n.

A5.2 FINITE DIFFERENCE EQUATIONS

These recurrences,

$$x(t+1) - x(t) = g(t) \qquad \text{EQ 1}$$

in which t is a continuous variable, were known to Euler, Bernoulli, Lagrange, and Laplace, and have been the object of constant study ever since. The linear case is the simplest. Two difficulties have limited the results from the beginning:

1. The use of Bernoulli polynomials, which do not always converge.

2. The impossibility of distinguishing the properties characterizing every solution of the recurrence equation from those introduced by the arbitrary initial function on the interval (0, 1).

The first difficulty was remedied by Nielsen, who introduced polynomials converging uniformly and absolutely when the function on the right is an entire function of genus 1 or more. His work was followed by contributions by Guichard, Appell, and Hurwicz. Replacing the Bernoulli polynomials by carefully chosen entire functions, these workers created a general method which assures the convergence of a series to a solution in the case in which g is an arbitrary entire function. [For a review of this subject, see A. N. Sharkovskii, Yu. L. Maistrenko, and F. Yu. Romanenko, *Difference Equations and their Applications*, Kluwer, 1993.]

A5.3 FUNCTIONAL EQUATIONS

Around 1870, Schröder [S3][1] studied the equation,

$$\varphi[f(x)] = S\varphi(x), \ f(0) = 0, \ 0 < S < 1 \qquad \text{EQ 2}$$

in which x is a real variable, φ is the unknown function, and f is an analytic function of x. This functional equation is linked to the problem of iteration by the consideration of the recurrence

$$x_{n+1} = f(x_n), f'(0) = S$$

which is such that x_n / S^n tends, for $|x|$ sufficiently small, to an analytic function φ solving (EQ 2). A characteristic of this equation is that its solution provides a nonlinear change of variable $x = \varphi(x)$ such that the functional relation $y = f(x)$ is linearized in the form $Y = SX$. In effect, one has $Y = \varphi(y)$ and $\varphi(y) = S\varphi(x)$, where $\varphi[f(x)] = S\varphi(x)$.

Equation (2) was the point of departure of a series of researches on iteration problems during the period 1820 – 1900, including:

---

1. The codes in brackets refer to the Historical Bibliography at the end of this appendix and are identical to those in the bibliography of GM1.

- Koenigs, 1883 – 1885 [K8K13]

- Grevy, 1892 – 1897 [G7G9]

- Leau, 1895 – 1898 [L14L16]

- Lemeray, 1895 – 1899 [L7L25].

These investigations were devoted mainly to the case of a single real or complex variable, which is supposed to be sufficiently near to an asymptotically stable fixed point or periodic cycle.

A5.4 POINCARÉ

The work of Poincaré [P9] (1885) on periodic solutions of differential equations was destined to have the greatest significance, and to be the basis of numerous results obtained up to the present day.

Poincaré's fundamental idea was to reduce the study of an autonomous differential equation of the second order to the study of a transformation of a line into itself. First he interpreted the second order differential equation as a first order dynamical system in the plane. Then, he drew a *section*, a curve "without contact" with the trajectories, and followed the trajectories (if recurrrent) to their first return across this section, defining the transformation. He also considered a transformation of a circle into itself, defined by trajectories on a torus (see [P9] para. 1.1.2.d for this case). In brief, this construction led him to a reduction in the dimension of the problem, and to the notion of a *limit cycle*. His interest in this method for studing the qualitative properties of solutions, and in particuliar, problems of bifurcation, led the great mathematician very naturally to generalize his method to dynamical systems in three dimensions, reducing these problems to the study of a transformation of a surface into itself (see [P9], especially para. 1.1.2.e, and [P10]). This permitted him, then, to develop a certain analogy between the fixed points of a transformation of the plane, and the singular points of a differential equation of second order, with behaviors of saddle, focus, and nodal type [P9]. At the same time, he defined the invariant curves passing through the saddle, by means of two convergent series.

This is the idea behind the iteration tool, and its traditional application mentioned at the beginning of this appendix. It opened a way to the solution of various problems attached to the discovery of periodic solutions, with regard to which Poincaré said,

> these yield us the solutions so precious, that is to say, they are the only breach through which we can penetrate into a place which up to now has been reputed to be inaccessable.

This breach was then to open a bit more with the restricted problem of three bodies, the first application to physical systems. This problem, a conservative, autonomous, dynamical system with two degrees of freedom, is reduced to three dimensions through the intermediary of a known integral introduced by Euler. Poincaré [P9 – P10] was the first to make a further reduction to a plane endomorphism, a transformation from a surface of section into itself.[1] The same reasoning leads him to reduce a Hamiltonian differential equation of second order (with time-periodic Hamiltonian function) to a plane endomorphism, facilitating the study of trajectories near a periodic motion. The connection between these two problems follows from the fact that the equations of the dynamical system (conservative, autonomous, with two degrees of freedom) can be reduced in the neighborhood of a periodic trajectory to a differential equation of second order with a timeperiodic Hamiltonian function. The three volumes of his celebrated work, *Méthodes nouvelles de la Mécanique Céleste* [P10], are primarily devoted to these questions, due to their fundamental focus on astronomy.

This same threebody problem led Poincaré to establish, through a more refined study of the invariant curves passing through a saddle point of a plane endomorphism, the notion of *doubly asymptotic point* (*homoclinic* or *heteroclinic*) and to notice the extreme complexity of the figure obtained in the surface, which he described as

---

1. We use the word endomorphism in the way common in topology: a mapping from a space to itself. This includes both the invertible and noninvertible cases. In the context of Poincaré, the endomorphism is not only invertible, it is a diffeomorphism. See the introduction to Appendix 2 and A2.1 for the full definitions.

an inextricable tangle of curves. The importance of these notions for dynamical systems theory became evident later. They are directly related to the chaoticity of certain physical systems, and it is not surprising to see, since 1968, an explosive increase of publications on this subject.

A5.5 INDEPENDENT CONTEMPORARIES OF POINCARÉ

The publications of Poincaré since 1880 stimulated many papers on the problems of recurrence, iteration, and endomorphisms. From 1900 to 1920, the following most significant contributions may be quoted:

- Hadamard [H2 – H4], who occupied himself particularly with the determination of invariant curves passing through a fixed point of a planar endomorphism, not assuming analyticity as in Poincaré;

- LeviCivita [L24], who was interested in the stability of periodic solutions of dynamical systems, in the case in which a fixed point of a planar endomorphism has characteristic multipliers which are roots of unity;

- Cigala [C3], who attacked the same problem, when the characteristic multipliers are on the unit circle, but are not roots of unity;

- Lattès [L3 – L13] who studied iterations with several variables in the neighborhood of a fixed point or a periodic cycle in connection with the work of Schröder in higher dimensions, with and without analyticity;

- Julia [J1 – J22] and Fatou [F1 – F4] began a series of researches on the global behavior of iterated functions of one complex variable, and were the first to define the properties of the boundaries of the basin of an attractive fixed point which need not be simply connected [including the properties later called fractal, which have become so popular since 1985];

- Brouwer [B38 – B39], who studied the existence of fixed points of a continuous planar endomorphism;

- Bennett [B14 – B18], who treated the iteration of functions of several variables; and

- Cotton [C9], who extended the characteristic number of Liapounov [L33] to recurrences.

Each of these works carried a stone to the foundations of dynamical systems theory. It does not diminish their merits, however, to affirm that the base of that theory was founded by just a part of the scientific work of Poincaré, who truly opened an immense domain.

A5.6 BIRKHOFF

The first to exploit this domain was the American, G. D. Birkhoff [B24 – B32]. Although the product of an American university, he was the intellectual disciple of Poincaré, without doubt the greatest of these. According to Marston Morse, "Poincaré was the true teacher of Birkhoff." His first work concerning the study of dynamical systems, which began at the point attained by Poincaré at his death in 1912, perfectly illustrates this point. Recall that, in his last publication, Poincaré showed that the existence of periodic solutions in the restricted problem of three bodies may be deduced from a theorem on planar endomorphisms, for which he was able to give a proof only for very particular cases [P10]. This proof, which resisted Poincaré, was fournished by his disciple in January 1913, in a rigorous but simple form [B32].

The breach opened by Poincaré was to be considerably enlarged by Birkhoff with his

- introduction of the notion of *recurrent motion* [B29];

- idea of *transitivity* [B32];

- celebrated *ergodic theorem* [B32];

- major studies of hyperbolic fixed points (*saddles*) having heteroclinic and homoclinic points, and of stable and unstable elliptic fixed points (*centers*) [B28 – B29];

- concept of the *signature* for the qualitative description of a dynamical system [B32];

- contributions to the threebody problem [B32];

and many other subjects. Like Poincaré, he was able to provide ample problems for his followers interested in endomorphisms and dynamical systems.

A5.7 DENJOY

In France, in parallel with this work of Birkhoff, Arnaud Denjoy in 1932 took up a question left hanging in the work of Poincaré. This was the problem of the curves defined by differential equations on the surface of a torus, for which *circle maps* are the basic tool [D5]. Later (1958 – 1966) he was able to extend this work to tori of dimension higher than three [D6].

A5.8 THE RUSSIAN SCHOOL

Most of Birkhoff's extensions of the work of Poincaré concerned conservative systems. In its turn, the theory of nonlinear oscillations in the nonconservative case furnished a vast field of applications for the methods elaborated by Poincaré. This direction was exploited by the Russian school of nonlinear mechanics, and was the source of a great number of particularly important works on this subject.

Andronov was the first to recognize that the Van der Pol oscillator was an example of the *limit cycle* concept introduced by Poincaré [A8]. He showed how a piecewiselinear approximation to the nonlinear characteristic function of an electronic tube led to a onedimensional map establishing the emergence of a stable limit cycle, and the behavior of its transitional regime [A14]. He extended this method to the escapement mechanism of clocks, to multivibrators, and to control relays. In the latter case he analysed the phenomenon of the *sliding regime*. These applications occupied most of his famous book with Witt and Khaïkin of 1947 [A14]. The introduction of one-dimensional maps, in itself, enabled the elucidation of a number of bifurcations in the behavior of the nonlinear systems, and certain points in the theory of *structural stability* elaborated with Pontryagin [A10].

Much like Poincaré, Andronov was then led to use planar maps to study the behavior of systems described by autonomous differential equations of order three. Throughout, he considered systems with piece-wiselinear nonlinearities. Thus he was led from 1944 to 1947 to study a series of problems related to the stabilization of an airplane by an automatic pilot (problems posed by Mises and by Vishnegradsky) in collaboration with N. N. Bautin, A. G. Mayer, and G. S. Gorelik, and to other problems in the theory of nonlinear oscillations [A8 – A22].

The death of Andronov left, among his disciples, an important school of thought in the domains which he had opened. This school is known as the *Gorki (Andronov, or MandelsthamAndronov) school*. It has been particularly active since 1950, producing a great number of publications on the properties of recurrent solutions, applications to some physical systems, and to the study of differential equations to which it is possible to associate a map. More precisely, the contributions of the Gorki school include ideas on

- the critical case of stability [A19];

- the bifurcations under variation of parameters [A16][A20];

and applications of these ideas to differential equations, such as studies of

- the effect of parameters on periodic solutions [N12];

- the solutions of piecewiselinear equations [B37];

- the method of small parameters of Poincaré, from the point of view of mappings, with extension to discontinuous equations [N10];

- the method of averaging [N11]; and

- the application of all of these ideas to a great number of electrical and mechanical systems [A4 – A6] [B22] [B26].

Regarding the more recent activities of this school, Neïmark is among the most published [N1 – N18]. Since 1967, there has been an evolution of this literature toward problems of invariant manifolds of endomorphisms of spaces of arbitrary dimension. However, recent times show also an orientation toward the study of dynamical systems with homoclinic structures, in association with the chaotic behavior of certain solutions. This evolution, characterized by the desire to utilize and adapt the results of Poincaré and Birkhoff to the solution of problems other than those of celestial mechanics, is very clear in the later works of

- Neïmark [N18];

- Beljustina [B12];

- Shilnikov [S12 – S27];

- Mel'nikov [M12 – M13]; and

- Gavrilov [G3].

A5.9   THE JAPANESE SCHOOL

The same tendency is found in Japan in the works of Hayashi [H17 – H18], who used mappings to study the solutions of secondorder nonautonomous differential equations. He also made apparent, in the case of the equations of Duffing and Van der Pol, the homoclinic and heteroclinic structures of certain solutions (see also the works of Kawakami [K2]).

A5.10 CONSERVATIVE SYSTEMS

Studies such as those mentioned above, on dynamical systems with homoclinic structures, mainly concern nonconservative systems. The conservative (and thus also areapreserving) case, which relates most directly to the work of Birkhoff, has been the object since 1954 of more general researches. Among these we may cite

- Kolmogorov [K14 – K15];
- Siegel;
- Moser [M47 – M49];
- Arnold [A25 – A28];
- Anosov [A23 – A24];
- Sinaï [S17 – S18];
- Alekseev [A6];
- Smale [S31];
- Mel'nikov [M12 – M13];
- Reeb [R2]; and
- Halmos,

who have all introduced new ideas, and who have followed esssentially the goals of pure mathematics.

Thus, Kolmogorov [K14] and Arnold [A27] have shown that, for analytic dynamical systems, small perturbations of the Hamiltonian conserve many of the nearly periodic solutions on invariant tori, and that these tori are only deformed. Moser showed then that one has the same situation without the hypothesis of analyticity, in the context of area-preserving endomorphisms of an annulus. Some conditions of stability for a center, in the case of two dimensions, have been given by Arnold [A28] and by Moser [M48 – M49].

Within this body of work there have emerged the notions of *mixing, invariant spectra, discrete spectrum*, and *entropy* [A27], destined to clarify the stochastic properties of such systems. One class of dynamical systems with essentially stochastic properties, the *C systems*, has been the object of many studies [S17 – S18][A27]. This class includes geodesic flows on manifolds of negative curvature.

A5.11 THE AMERICAN SCHOOL

Parallel to this work, the American school has developed its research on flows and diffeomorphisms based on concepts of generic properties, structural stability, the nonwandering set of Birkhoff, and on different types of attractors, in the works of Peixoto [P16], Smale [S30S31], Palis [P17], Pugh [P18], Guckenheimer [G48], and Sotomayor [S20]. Earlier, in a related context, Thom [T4 – T5] was led to his work on the structural stability and singularities of smooth mappings and, based on these, his theory of catastrophes. This concerns the bifurcations of gradient dynamical systems defined by a potential function.

A5.12 NUMERICAL METHODS AND APPLIED WORK

With the appearance of computers of high speed, capacity, and precision, this pure mathematical research was joined by a significant number of works focused on concrete problems. The use of computers had made it possible to approach these problems by means of numerical solution of the equations of the systems. Two scientific areas provided fields of application: the physics of high energy particle accelerators, and astronomy. Since 1968, despite the problems of numerical sensitivity due, for example, to roundoff errors and to the conservative nature of the equations, a number of authors ([L20] [H20] [B8] [F16] [M56] [F17] [G31] [G34] [G35]) were able to materialize some theoretical results already known, and even, in certain cases, to discover new phenomena. Furthermore, the problems of particle accelerators and astronomy were subjected also to analytic methods. Certain results relating to this body of work were described in GM2.

A5.13 ITERATION THEORY

Returning to the point of view of recurrence, iteration, and difference equations, the works Norlund [N20], Montel [M46], and

Kuzma [K20] must be noted. Iteration theory, after the early works of Lattès, Julia, and Fatou, produced a very great number of works, including

- Montel [M45] on maps of several variables;

- Baker on maps of one complex variable, especially with characteristic multiplier equal to 1;

- Levy [L31] on iteration of the exponential function;

- Myrberg [M50 – M55] on real polynomials;

- Kuczma [K17 – K22] on continuous iterations, commuting functions, and convergence of the iterations;

- Barna [B6];

- Sharkovskii [S4 – S13] on first-order iterations with one real variable, and behavior near an attractor;

- Pulkin [P14] on firstorder iterations with one real variable, and stochastic invariant sets;

- Ulewicz [U1] on conditions for stability of a fixed point of a firstorder iteration when the function is only differentiable at that point;

- Drewniak [D11] on the geometric convergence of an iterated sequence in the firstorder case, without assuming differentiability at the fixed point; and

- Nishino and Yoshioka [N19], extending the results of Julia to the case of two complex variables.

Regarding fractional iterations, we note the contributions of Levy [L31], Szekeres, Bödewadt, Kuczma, and Smado. As for the Schröder equation, it has continued to be the target of numerous works by Kuczma, Urabe, Targonski, and others. After Appel, the perspective of difference equations occasioned the contributions of Birkhoff, Perron, Carmichael, Norlund [N20], Raclis, Bochner, Ghermanesko [G5], Ta Li, and Harris and Sibuya [H13 – H16]. For all these, see the very complete bibliography of [K20].

A5.14 THE METHODS OF LIAPUNOV

Since 1958, the extension of the second method of Liapunov [L33] (concerning the stability of the solutions of differential equations) to recurrences appeared in the work of Hahn [H5 – H10], who was the first to announce these results. The first method of Liapunov (that of the characteristic numbers) had already been extended to recurrences by Cotton, as mentioned above. This objective was taken up again by Kalman [K1] and other authors, including Fischer [F14 – F15], who made a generalization to locally compact, separated spaces. The method of characteristic numbers of Liapunov was developed by Demidovitch [D3 – D4] and Panov [P1 – P7] in the case of nonautonomous systems of recurrences. Panov was interested also in the classification of singular points in Euclidean spaces of several dimensions, and in the behavior of solutions near these points.

A5.15 PERIODIC SOLUTIONS

As for problems tied to the periodic solutions of nonlinear recurrences, we find them in works by Tsypkin [T10], Jury [J24], Urabe [U2], and Halanay [H12]. Halanay was also interested in the case of almost periodic solutions. In another direction, the problem of invariant manifolds of recurrences took the attention of Haag [H1], Urabe [U2], Neïmark [N16], and Halanay [H12].

A5.16 CONTROL THEORY

Since 1956, the development of systems controlled by sampled data and by impulses, in the context of automation, has led to researches concerning nonlinear recurrences. These are very numerous; we will particularly cite those of

• Jury [J23], Clark [C4] (USA);

- Tsypkin [T9T10], Kuntsévitch [K23], Tchekovoï [T7] (Russia);

- Kodata, Kodama (Japan);

- Wegrzyn (Poland);

- Vidal [V1] (France).

See also [B33], [B41].

A5.17 OTHER APPLICATIONS

In economics, around 1950, Samuelson [S2] had formulated models in the form of implicit nonlinear recurrences and studied the solutions near a stable fixed point. The interest in this type of equation by various scholars in other fields — biology, ecology (May [M3 – M6] and [M57]), chemical kinetics (Rössler [R3 – R7]), physics (Pomeau, Gervois, Derida [D7]) — constitutes a very recent and broad phenomenon. This is based on the fact that the simplest onedimensional endomorphism, defined by a function with a single extremum, has solutions with complex dynamical behavior, suitable for describing natural phenomena, such as chaos in biology. In the domain of pure mathematics, since 1975 these simplest maps have been the subject of an increasing number of publications, from the point of view of bifurcation theory as well as that of ensemble dynamical properties such as invariant measure and entropy. See, for example, [C7], [C8], [C10], [C11], [D7], [G10], [G53], [M14], [M4], [M39], and [Z2].

A5.18 CONCLUSION

Finally, two recent works, those of Neïmark (1972) [N1] and Gauchus (1976) [G2], confirm the increasing role of endomorphisms, relative to differential equations, in the interdisciplinary work being treated with nonlinear dynamical systems and their applications. This tour of the frontier will be incomplete without mentioning the contributions of the members of the Lefshetz Institute of Dynamical Systems in the USA, which for a long time has directed a part of its activities to the domain of the life sciences, along with numerous other teams, all over the world, working in the same direction.

A5.19 HISTORICAL BIBLIOGRAPHY

[A 4] A.S. ALEKSEEV. Application des transformations ponctuelles a l'etude des systemes dynamiques non lineaires a impulsions. *Radiofisica*. t. 13. no 8. 1970.

[A 5] A.S. ALEKSEEV. Etude de la dynamique des systemes non lineaires a impulsions avec partie commandee a parametres repartis. *Radiofisica*. t. 13. no 8. 1970.

[A 6] V.M. ALEKSEEV. Systemes dynamiques quasi-aleatoires. *Mat. Shornik*. t. 76. 118. no 1. 1968: t. 77. 119 no 4. 1968. t. 78. 120 no 1. 1969.

[A 8] A.A. ANDRONOV. Cycles limites de Poincare et theorie des oscillations. *Dokl. Akad. Nauk. SSSR* VI. 1928 et *Oeuvres completes d'Andronov*. Edition de l'Acad. Sc. d'U.R.S.S. 1956.

[A 9] A.A. ANDRONOV. A.A. WITT. Solutions periodiques discontinues et theorie du multivibrateur d'Abraham-Bloch. *Dokl. Akad. Nauk*. no 8. 1930. p. 189 et *Oeuvres completes d'Andronov*. Edition de l'Acad. Sc. d'U.R.S.S. 1956. p. 65.

[A 10] A.A. ANDRONOV. L.S. PONTRYAGIN. Systemes grossiers. *Dokl. Akad. Nauk*. vol. 14. 1937. p. 247 et *Oeuvres completes d'Andronov*. Edition de l'Acad. Sc. d'U.R.S.S. 1956. p. 181.

[A 11] A.A. ANDRONOV. Mandelshtam et la theorie des oscillations non lineaires. *Oeuvres completes d'Andronov*. Edition de l'Acad. Sc. d'U.R.S.S. 1956. pp. 441 – 472.

[A 12] A.A. ANDRONOV. Theorie des transformations ponctuelles de Poincaré-Brouwer-Birkhoff et theorie des oscillations non lineaires. *Vestnik Akad. Nauk. SSSR*. no 6. 1944.

[A 13] A.A. ANDRONOV. A.G. MAYER. Le probleme de Mises dans la theorie de la regulation directe et la theorie des transformations ponctuelles du plan. *Dokl. Akad. Nauk*. SSSR. T. 43. no 2. 1944.

[A 14] A.A. ANDRONOV. A.A. WITT. S. KHAIKIN. *Theory of oscillators*. Pergamon Press. 1966.

[A 15] A.A. ANDRONOV. E.A. LEONTOVICH. Theorie des changements de comportement qualitatif des trajectoires du plan de phase. *Dokl. Acak. Nauk. SSSR*. vol. 21. no 9. 1938.

[A 16] E.A. ANDRONOVA-LEONTOVICH. L.N. BELIUS-TINA. Theorie des bifurcations des systemes dynamiques du deuxieme ordre et son application a l'etude des problemes non lineaires da la theorie des oscillations. *Travaux du Symposium International sur les oscillations non lineaires. Kiev. Septembre 1961. Actes*. vol. 2 (1963). pp. 728.

[A 17] A.A. ANDRONOV. N.N. BAUTIN. G.S. GORELIK. Theorie de la regulation indirecte avec frottement sec. *Automatika i Telemec*. t. 7. no 1. 1946.

[A 18] A.A. ANDRONOV. A.G. MAYER. Le probleme de Vichnegradski dans la theorie de la regulation directe. *Automatika i Telemec*. t. 8. no 5. 1947. et t. 14. no 5. 1953.

[A 19] A.A. ANDRONOV. E.A. LEONTOVITCH. I.I. GORDON. A.G. MAYER. *Theorie qualitative des systemes dynamiques du deuxieme ordre*. Nauka. Moscou. 1966.

[A 20] A.A. ANDRONOV. E.A. LEONTOVITCH. I.I. GORDON. A.G. MAYER. *Theorie des bifurcations des systemes dynamiques dans le plan*. Nauka. Moscou. 1967.

[A 21] A.A. ANDRONOV. N.N. BAUTIN. Theorie de la stabilisation du vol d'un avion neutre a l'aide d'un pilote automatique. *Izv. Akad. Nauk. SSSR*. no 3. 1955; no 6. 1955.

[A 22] A.A. ANDRONOV. N.N. BAUTIN. G. GORELIK. Auto-oscillations d'un schema simplifie avec helice a pas variable automatique. *Dokl. Akad. Nauk. SSSR*. t. 47. no 4. 1945 (en francais).

[A 23] D.V. ANOSOV Flux geodesique sur des varietes dermees de Riemann avec courbure negative. *Travaux de l'Institut Mathematique. V.A. Steklov*. t. 90. Nauka Moscou. 1967.

[A 24] D.V. ANOSOV Ya. G. SINAI. Systemes ergodiques differentiables. *Uspeki Mat. Nauk*. t. 22. no 5. (137). 1967.

[A 25] V.I. ARNOLD. Petits diviseurs. Transformation d'un cercle en luimeme. *Izv. Akad. Nauk. SSSR*. Mat. t. 25. no 2. 1961.

[A 26] V.I. ARNOLD. Petits diviseurs et problemes de stabilite du mouvement en mecanique classique et celeste. *Uspeki Mat. Nauk.* t. 18. no 6. 1963.

[A 27] V.I. ARNOLD. A. AVEZ. *Problemes ergodiques de la mecanique.* GauthierVillars. Paris. 1966.

[A 28] V.I. ARNOLD. Singularites des transformations differentiables. *Uspeki Mat. Nauk.* t. 23. no 1. 1968. pp. 3 – 44.

[A 29] V.I. ARNOLD. Instabilite des systemes dynamiques a plusieurs degres de liberte. *Dokl. Akad. Nauk. SSSR.* t. 156. no 1. 1964. pp. 9 – 12.

[B 6] B. BARNA. Uber die Iteration reeler Funktionen. I.II. *Publ. Math. Debrecen* 2 (1951). pp. 5063.

[B 12] L.N. BELIOUSTINA. V.N. BELIUK. Systeme non autonome d'equations de phase avec petit parametre. contenant un tore invariant. et courbes homoclines. *Radiofisica.* t. 15. no 7. 1972.

[B 14] A.A. BENNET. The iteration of functions of one variable. *Bull. Amer. Math. Soc.* 22 (1915). p. 12.

[B 15] A.A. BENNET. The iteration of functions of one variable. *Ann. of Math.* (2) 17 (1916). pp. 23 – 60.

[B 16] A.A. BENNET. A case of iteration in several variables. *Ann. of Math.* (2) 17 (1916). pp. 188 – 196.

[B 17] A.A. BENNET. Note of the preceding paper. *Ann. of Math.* (2) 17 (1916). p. 123.

[B 18] A.A. BENNET. A case of iteration in several variables. *Bull. Amer. Math. Soc.* 22 (1916). pp. 487 – 488.

[B 22] L.V. BESPALOVA. Theorie du mecanisme vibropercutant. *Izv. Akad. Nauk. SSSR.* no 5. 1957.

[B 24] G.D. BIRKHOFF. The generalized Riemann problem from linear differential equations and the allied problems for linear difference and qdifference equations. *Bull. Amer. Math. Soc.* 19 (1913). pp. 508 – 509.

[B 25] G.D. BIRKHOFF. The generalized Riemann problem for linear differential equations and the allied problems for linear difference and qdifference equations. *Proc. Amer. Acad. Arts Sci.* 49 (1930). pp. 521568.

[B 26] G.D. BIRKHOFF. Dynamical systems with two degrees of freedom. *Proc. Nat. Acad. Sci. U.S.A.* 3 (1916). pp. 314 – 316.

[B 27] G.D. BIRKHOFF. Dynamical systems with two degrees of freedom. *Trans. Amer. Math. Soc.* 18 (1917). pp. 199 – 300.

[B 28] G.D. BIRKHOFF. Surface transformations and their dynamical applications. *Acta Math.* 43 (1922). pp. 1 – 119.

[B 29] G.D. BIRKHOFF. Nouvelles reherches sur les Systemes Dynamiques. *Memoriae Pont. Acad. Sci. Novi Lyncae.* S. 3. vol. 1. 1935. pp. 85 – 216.

[B 30] G.D. BIRKHOFF. *Dynamical Systems.* Am. Math. Soc. Colloquium Publications. vol. IX. 1966.

[B 31] G.D. BIRKHOFF. P.A. SMITH. Structure analysis of surface transformations. *Journal de Math.* S. 9. vol. 7. 1926. pp. 345 – 379.

[B 32] G.D. BIRKHOFF. *Collected mathematical papers.* Dover Pub. Inc. N.Y.. 1968.

[B 33] P. BORNE. J.C. GENTINA. F. LAURENT. Sur une etude par majoration de la stabilite des suites recurrentes vectorielles non lineaires. *Transformations Ponctuelles et Applications. (Colloque CNRS Sept. 1973).* Editions CNRS. Paris 1976.

[B 37] V.A. BROUSIN. Yu. I. NEIMARK. M.I. FEIGIN. Dependance par rapport a des parametres des regimes periodiques des systemes a relais. *Radiofisica.* t. 6. no 4. 1963.

[B 38] L.E.J. BROUWER. Aufzahlung der periodischen Transformationen des Torus. *K. Akad. van Wetenschappen Amsterdam.* vol. 21. 1919. pp. 1352 – 1356.

[B 39] L.E.J. BROUWER. Continuous oneone transformations of surfaces in themselves. *K. Akad. van Wetenschappen Amsterdam.* vol. 11. 1909. pp. 788 – 798; vol. 12. 1909. pp. 286 – 297; vol. 13. 1910. pp. 767 – 777; vol. 14. 1911. pp. 300 – 310; vol. 15. 1912. pp. 352 – 360.

[B 41] Y. BURIAN. R. BOTTURA. Controle por comutacao do motor serie. *Iro Simposio Brasileiro de Ciencas Mecanicas.* Campinas 1973.

[C 3] A.R. CIGALA. Sopra un criterio di instabilita. *Annali di Matematica.* Ser. 3. t. 11. 1905.

[C 4] R.N. CLARK. FRANKLIN. Limit cycles in pulse modulated systems. *Journal of Spacecrafts and Rockets.* 69.

[C 7] M. COSNARD. Sur les suites iteratives oscillantes: demonstration d'un theoreme de Pulin. *Colloque sur les Processus Iteratifs. La Garde Freinet.* Mai 1979.

[C 8] M. COSNARD. M. DELARCHE. A. EBERHARD. H. LEPELTIER. *The three possible behaviours for the iterates sequence of a real continuous function.* Rapport no 89. 1977. Laboratoire associe CNRS no 7. Grenoble.

[C 9] E. COTTON. Sur la notion de nombre caracteristique de M. Liapunof. *Ann. Ecole Norm.* 3. no 36. Juin 1919. pp. 128 – 185.

[C 10] J. COUOT. Computer simulation of invariant measures of a discrete dynamic system possessing chaotic regimes. *Congres Informatica 77.* Bled (Yougoslavie). 1977.

[C 11] J. COUOT. C. GILLOT. Equations fonctionnelles des densites de mesures invariantes par un endomorphisme de [0.1] et simulation numerique. *Congres Equadiff 78. Florence. mai 1978. Actes*, p. 139.

[D 3] V.B. DEMIDOVITCH. Systemes lineaires d'equations aux differences. *Different. uravn.* t. 7. no 5. 1971.

[D 4] V.B. DEMIDOVITCH. Sur un indice de stabilite des equations aux differences. *Different. uravn.* t. 5. no 7. 1969.

[D 5] A. DENJOY. Sur les courbes definies par des equations differentielles a la surface du tore. *Journ. de Math.* t. 11. fasc. 4. 1932. pp. 333 – 375.

[D 6] A DENJOY. *C.R. Acad. Sc. Paris.* t. 247. p. 1072; t. 247. p. 1096; t. 247. p. 1923; t. 248. p. 28; t. 248. p. 325; t. 248. p. 497; t. 248. p. 1253; t. 261. p. 2917; t. 261. p. 4293; t. 261. p. 4579; t. 263. p. 67.

[D 7] B. DERRIDA. A. GERVOIS. Y. POMMEAU. Iteration d'endomorphismes de la droite reelle et representation des nombres. *C.R. Acad. Sc. Paris.* t. 285. A. 1977. pp. 4 – 46.

[D 11] J. DREWNIAK. Convergence geometrique des suites recurrentes. *Transformations Ponctuelles et Applications. (Colloque CNRS. Sept. 73. Toulouse).* Editions CNRS Paris 1976.

[F 1] P. FATOU. Memoire sur les equations fonctionnelles. *Bull. Soc. Math. France.* 47 (1919). pp. 161 – 271. 48 (1920). pp. 33 – 94, 208 – 314.

[F 2] P. FATOU. Sur les domaines d'existence de certaines fonctions uniformes. *C.R. Acad. Sc. Paris.* 173 (1921). pp. 344 – 346.

[F 3] P. FATOU. Sur les fonctions qui admettent plusieurs theoremes de multiplication. *C.R. Acad. Sc. Paris* 173 (1921). pp. 571 – 573.

[F 4] P. FATOU. Sur un groupe de substitutions algebriques. *C.R. Acad. Sc. Paris* 173 (1921). pp. 694 – 696.

[F 14] P. FISCHER. Sur une extension de la deuxieme methode de Liapunov aux applications. *C.R. Acad. Sc. Paris.* t. 270. pp. 235 – 238. Serie A. 1970.

[F 15] P. FISCHER. Sur une extension de la deuxieme methode de Liapunov aux recurrences non autonomes. *C.R. Acad. Sc. Paris.* t. 270. pp. 762 – 765. Serie A. 1970.

[G 2] E.V. GAUCHUS. *Etude des systemes dynamiques: Methode des transformations ponctuelles.* Nauka Moscou. 1976.

[G 3] N.K. GAVRILOV. L.P. SHILNIKOV. Systemes dynamiques tri-dimensionnels proches d'un systeme aveo courbes homoclines non grossieres. *Mat. Sbornik.* t. 88. no 4. 1972: t. 90. no. 1. 1973.

[G 5] M. GHERMANESCO. Sur les equations aux differences finies. *Annali di Mat.* Ser. IV. t. 12. 1934.

[G 7] A. GREVY. Sur les equations fonctionnelles. *Bull. Sci. Math.* (2) 16 (1892). pp. 311 – 313.

[G 8] A. GREVY. Etude sur les equations fonctionnelles. *Ann. Sci. Ecole Norm. Sup.* (3) 11 (1894). pp. 249323; 13 (1896). pp. 295 – 338.

[G 9] A. GREVY. Equations fonctionnelles avec second membre. *Bull. Soc. Math. France.* 25 (1897). pp. 57 – 63.

[G 10] S. GROSSMANN. S. THOMAE. Invariant distributions and stationary correlations functions of onedimensional discrete processes. *Z. Naturforsch.* 32a. 1977. pp. 1353 – 1363.

[G 11] I. GUMOWSKI. C. MIRA. Sur une solution particuliere de l'equation de Schröder. *C.R. Acad. Sc. Paris.* 259 (1964). p. 2952.

[G 12] I. GUMOWSKI. C. MIRA. Sur une solution particuliere explicite de l'equation fonctionnelle de Schröder. *C.R. Acad. Sc. Paris.* 259 (1964). p. 4476.

[G 13] I. GUMOWSKI. C. MIRA. Sur un algorithme de determination du domaine de stabilite d'un point double d'une recurrence non lineaire du deuxieme ordre a variables reelles. *C.R. Acad. Sc. Paris.* 260 (1965). p. 6524.

[G 14] I. GUMOWSKI. C. MIRA. Determination graphique de la frontiere de stabilite d'un point d'equilibre d'une recurrence non lineaire du deuxieme ordre a variables reelles. Application au cas ou les seconds membres de la recurrence ne sont pas analytiques. *Electronics Letters* 2 (1966). no 5.

[G 15] I. GUMOWSKI. C. MIRA. Etude des points singuliers a l'infini d'une recurrence autonome du deuxieme ordre a variables reelles. *C.R. Acad. Sc. Paris.* 263 (1966). p. 547. Serie A.

[G 16] I. GUMOWSKI. C. MIRA. Construction de points doubles et de cycles d'une recurrence. Application a l'etude d'une bifurcation. *C.R. Acad. Sc. Paris* 263 (1966). p. 837. Serie A.

[G 17] I. GUMOWSKI. C. MIRA. Sur la formulation equivalente de Caratheodory relative a un probleme variationnel. *C.R. Acad. Sc. Paris.* 264 (1967). p. 259. Serie A.

[G 18] I. GUMOWSKI. C. MIRA. E. RIBERI. Sur la frontiere du domaine de stabilite d'un point double d'une recurrence non lineaire. Lorsque cette frontiere ne contient pas de point double. *C.R. Acad. Sc. Paris.* 265 (1967). p. 59. Serie A.

[G 19] I. GUMOWSKI. C. MIRA. Problemes de sensibilite lies a certaines bifurcations dans les recurrences non lineaires. *2nd IFAC Symposium on System Sensitivity and Adaptivity. Dubrovnik* (1968).

[G 20] I. GUMOWSKI. C. MIRA. Sensitivity problems related to certain bifurcations in nonlinear recurrence relations. *Automatica.* 5 (1969). p. 303317.

[G 21] I. GUMOWSKI. K.H. REICH. Mouvement synchrotonique dans un accelerateur a protons en presence de charge d'espace. *C.R. du 2eme Congress URSS sur les Accelerateurs des Particules Chargees.* Moscou (1970). *Nauka.* (1972). t. 2. p. 2328.

[G 22] I. GUMOWSKI. C. MIRA. *L'Optimisation La Theorie et ses Problemes.* Dunod (1970).

[G 23] I. GUMOWSKI. C. MIRA. Boundaries of stochasticity domains in Hamiltonian systems. *Proc. 8th Int. Conf. on High Energy Accelerators, CERN. Geneve (1971).* p. 374 – 376.

[G 24] I. GUMOWSKI. Determination of an equipment design criterion for the Cern PS Booster from stability considerations of coherent synchrotron motion. *Proc. 8th Int. Conf. on High Energy Accelerators. CERN. Geneve (1971).* pp. 360 – 362.

[G 25] I. GUMOWSKI. C. MIRA. Sur la distribution des cycles d'une recurrence ou transformation ponctuelle conservative du deuxieme ordre. *C.R. Acad. Sc. Paris.* 274 (1972). p. 1271. Serie A.

[G 26] I. GUMOWSKI. J.K. TRICKETT. Sur les bifurcations liees aux cas d'exception d'une recurrence conservative du deuxieme ordre. *C.R. Acad. Sc. Paris.* 275 (1972). p. 147. Serie A.

[G 27] I. GUMOWSKI. C. MIRA. Sur la structure des lignes invariantes d'une recurrence ou transformation ponctuelle conservative du deuxieme ordre. *C.R. Acad. Sc. Paris.* 275 (1972). p. 869. Serie A.

[G 28] I. GUMOWSKI. Sur les bifurcations liees aux cas d'exception degeneres d'une recurrence conservative du deuxieme ordre. *C.R. Acad. Sc. Paris.* 275 (1972). p. 939. Serie A.

[G 29] I. GUMOWSKI. C. MIRA. Stochastic solutions in a conservative dynamic system. *Proc. 6th Int. Conf. on Non-linear Oscillations, Poznan. (1972).* PWNWarszawa. (1973). pp. 417 – 438.

[G 30] I. GUMOWSKI. C. MIRA. Determination des courbes invariantes fermees d'une recurrence ou d'une transformation ponctuelle non lineaire. voisine d'une recurrence conservative lineaire. *C.R. Acad. Sc. Paris.* 276 (1973). p. 333. Serie A.

[G 31] I. GUMOWSKI. Some properties of large amplitude solutions of conservative dynamic systems. Part 1: Quadratic and cubic nonlinearities. Part 2: bounded nonlinearities. *Rapport CERN/SI/Int. BR/721*. Geneve. 1972.

[G 32] I. GUMOWSKI. K. SCHINDL. Periodic solutions of a second order phase lock system. *Proc. 7th AICA Congress, Prague. (1973)*. pp. 139 – 142.

[G 33] I. GUMOWSKI. Solution structure of a conservative second order recurrence with an unbounded nonlinearity. Colloque International CNRS. *Transformations Ponctuelles et leurs Applications. Toulouse. (1973)*. Editions CNRS. Paris. 1976.

[G 34] I. GUMOWSKI. J.K. TRICKETT. Some properties of a conservative second order recurrence with an unbounded nonlinearity. *Colloque International CNRS. Transformations Ponctuelles et leurs Applications, Toulouse. (1973)*. Editions CNRS. Paris 1976.

[G 35] I. GUMOWSKI. Stochastic effects in longitudinal phase space. *9th Int. Conf. on HighEnergy Accelerators. Stanford. (1974)*. Proc. pp. 439 – 443.

[G 36] I. GUMOWSKI. C. MIRA. Point sequences generated by twodimensational recurrences. *Proc. IFIP Congress 74. Stockholm. (1974)*. pp. 851 – 855.

[G 37] I. GUMOWSKI. C. MIRA. Bifurcation pour une recurrence du deuxieme ordre. par traversee d'un cas critique avec duex multiplicateurs complexes conjugues. *C.R. Acad. Sc. Paris. 278 (1974)*. p. 1591. Serie A.

[G 38] I. GUMOWSKI. C. MIRA. Sur les recurrences ou transformations ponctuelles du premier ordre avec inverse non unique. *C.R. Acad. Sc. Paris. 280 (1975)*. p. 905. Serie A.

[G 39] I. GUMOWSKI. Invariant curve structure of a second order conservative point mapping induced by a degenerate fixed point. *7th International Conference on Nonlinear Oscillations. Berlin. September 1975*.

[G 40] I. GUMOWSKI. C. MIRA. Accumulation de bifurcations dans une recurrence. *C.R. Acad. Sc. Paris. 281 (1975)*. p. 45. Serie A.

[G 41] Z. GUMOWSKI. I. GUMOWSKI. Random number generation and complexity in certain dynamic systems. *Proc. Sympo. Informatica 75 Bled. Yugoslavia. October 1975*. p. 3.4.1 – 4.

[G 42] I. GUMOWSKI. Contribution de l'automatique au probleme de reversibilite microscopique irreversibilite macroscopique. *RAIRO de l'AFCET.* 10(7). (1976). pp. 7 – 42.

[G 43] I. GUMOWSKI. C. MIRA. Boxwithinabox bifurcation structure and the phenomenon of chaos. *Informatica76. Bled. (1976).*

[G 44] I. GUMOWSKI. C. MIRA. Solutions chaotiques bornees d'une recurrence ou transformation ponctuelle du 2e ordre a inverse non unique. *C.R. Acad. Sc. Paris.* t. 285. Serie A (1977). pp. 477 – 480.

[G 45] I. GUMOWSKI. C. MIRA. Bifurcation destabilisant une solution chaotique d'un endomorphisme du 2e ordre. *C.R. Acad. Sc. Paris.* t. 286. serie A. (1978). pp. 427 – 431.

[G 46] I. GUMOWSKI. K.H. REICH. Synchrotron motion in the presence of space charge. *Rapport CERN SI/Int DL/706 (1970).*

[G 47] I. GUMOWSKI. Invariant curves of a second order recurrence possessing a nonclassical singularity. *8th International Conference on Nonlinear Oscillations.* Prague Sept. 1978.

[G 53] J. GUCKENHEIMER. Bifurcation of quadratic functions, in "Bifurcation theory and applications in scientific disciplines," *Ann. of N.Y. Acad. of Sc.* vol. 316. 1979. pp. 79 – 85.

[H 1] J. HAAG. Sur la stabilite des points invariants d'une transformation. *Bull. des Sc. Math.* t. 73. 1949. p. 123.

[H 2] J. HADAMARD. Sur les transformations ponctuelles. *Bull. Soc. Math. France.* 34. (1906). pp. 349 – 363.

[H 3] J. HADAMARD. Sur l'iteration et solutions asymptotiques des equations differentielles. *Bull. Soc. Math. France.* 29 (1901). pp. 224 – 228.

[H 4] J. HADAMARD. Two works on iteration and related questions. *Bull. Amer. Math. Soc.* 50 (1944). pp. 67 – 75.

[H 5] W. HAHN. Uber die Reduzibilitat einer speziellen geometrischen Differenzengleichung. *Math. Nachr.* 5 (1951). pp. 347 – 354.

[H 6] W. HAHN. Uber eneigentliche Losungen Linearer Geometrischer Differenzengleichungen. *Math. Ann.* 125 (1952). pp. 6781; *Berichtigung Math. Ann.* 125 (1953). p. 324.

[H 7] W. HAHN. Die mechanische Deutung einer geometrischen Differenzgleichung. *Z. angew. Math. Mech.* 33 (1953). pp. 270 – 272.

[H 8] W. HAHN. On the application of the matrix calculus in the theory of geometrical difference equations. *Proc. Indian Acad. Sci.* Sect. A 51 (1960). pp. 137 – 145.

[H 9] W. HAHN. *Stability of motion.* t. 138. Springer-Verlag Heidelberg. 1967.

[H 10] W. HAHN. *Theory and application of Liapunov's direct method.* Prentice Hall. 1963.

[H 12] A. HALANAY. D. WEXLER. *Teoria calitativa a sistemelor cu impulsuri.* Bucharest 1968.

[H 13] Jr. W.A. HARRIS. Y. SIBUYA. Asymptotic solutions of systems of nonlinear difference equations. *Arch. Rational Mech. Anal.* 15 (1964). pp. 377 – 395.

[H 14] Jr. W.A. HARRIS. Y. SIBUYA. Note on linear difference equations. *Bull. Amer. Math. Soc.* 70 (1964). pp. 123 – 127.

[H 15] Jr. W.A. HARRIS. Y. SIBUYA. General solution of nonlinear difference equations. *Trans. Amer. Math. Soc.* 115 (1965). pp. 62 – 75.

[H 16] Jr. W.A. HARRIS. Y. SIBUYA. On asymptotic solutions of systems of nonlinear difference equations. *J. reine angew. Math.* 222 (1966). pp. 120135.

[H 17] C. HAYASHI. Y. UEDA. Behaviour of solutions for certain types on non linear differential equations of the second order. VI International conference on nonlinear oscillations. Podznan Sept. 72. *Proceeding nonlinear vibration problems 1973,* 14. pp. 341 – 351.

[H 18] C. HAYASHI. Y. UEDA. H. KAWAKAMI. Transformation theory as applied to the solutions of nonlinear differential equations of 2nd order. *Int. J. Nonlinear Mechanics.* vol. 4. 1969. pp. 235 – 255.

[J 1] G. JULIA. Sur les substitutions rationnelles. *C.R. Acad. Sci. Paris.* 164 (1917). pp. 1098 – 1100.

[J 2] G. JULIA. Memoire sur l'iteration des fonctions rationnelles. *J. Math. Pures Appl..* t. 4. 1. (1918). (7eme serie). pp. 47 – 245.

[J 3] G. JULIA. Sur l'iteration de fractions rationnelles. *C.R. Acad. Sci. Paris.* 166 (1918). pp. 61 – 64.

[J 4] G. JULIA. Sur des problemes concernant l'iteration des fractions rationnelles. *C.R. Acad. Sci. Paris.* 166 (1918). pp. 153 – 156.

[J 5] G. JULIA. Sur les substitutions rationnelles. *C.R. Acad. Sci. Paris.* 166 (1918). pp. 599 – 601.

[J 6] G. JULIA. Sur quelques problemes relatifs a l'iteration de fractions rationnelles. *C.R. Acad. Sci. Paris.* 168 (1919). pp. 147 – 149.

[J 7] G. JULIA. Memoire relatif a l'etude des substitutions rationnelles a une variable. *Bull. Sci. Math.* (2) 43 (1919). pp. 106 – 109.

[J 8] G. JULIA. Sur la permutabilite des substitutions rationnelles. *C.R. Acad. Sci. Paris.* 173 (1921). pp. 690 – 693.

[J 9] G. JULIA. Memoire sur la permutabilite des fractions rationnelles. *Ann. Sci. Ecole Norm. Sup.* (3) 39 (1922). pp. 131 – 215.

[J 10] G. JULIA. Les equations fonctionnelles et la representation conforme. *C.R. Acad. Sci. Paris.* 174 (1922). pp. 517 – 519.

[J 11] G. JULIA. Sur la transformation de substitutions rationnelles en substitutions lineaires. *C.R. Acad. Sci. Paris.* 174 (1922). pp. 800 – 802.

[J 12] G. JULIA. Sur les substitutions rationnelles a deux variables. *C.R. Acad. Sci. Paris.* 175 (1922). pp. 1182 – 1185.

[J 13] G. JULIA. Series de fractions rationnelles d'iteration. *C.R. Acad Sci. Paris.* 180 (1925). pp. 563 – 566.

[J 14] G. JULIA. Sur les series de fonctions iterees. *C.R. Acad. Sci. Paris.* 181 (1925). pp. 1119 – 1122.

[J 15] G. JULIA. Les series d'iteration et les fonctions quasianalytiques. *C.R. Acad. Sci. Paris.* 180 (1925). pp. 720 – 723.

[J 16] G. JULIA. Sur la convergence des series de fractions rationnelles iterees. *C.R. Acad. Sci. Paris.* 191 (1930). pp. 987 – 989.

[J 17] G. JULIA. Les series de fractions rationnelles iterees et les fonctions quasianalytiques. *Arkiv fur Mat.* 22 A (1930). Nr 7 8 pp.

[J 18] G. JULIA. Memoire sur la convergence des series formes avec les iterees successives d'une fraction rationnelle. *Acta Math.* 56 (1930). pp. 149 – 195.

[J 19] G. JULIA. Fonctions continues sans derivees formees avec les iterees d'une fraction rationnelle. *Ann. Sci. Ecole Norm. Sup.* (3) 48 (1931). pp. 1 – 14.

[J 20] G. JULIA. Memoire sur l'extension du theorie d'Abel aux series d'iterees EanRn(z). *Ann. Sci. Ecole Norm. Sup.* (3) 48 (1931). pp. 439 – 495.

[J 21] G. JULIA. Sur l'allure des series d'iterees au voisinage des frontieres de convergence. *C.R. Acad. Sci. Paris.* 193 (1931). pp. 690 – 692.

[J 22] G. JULIA. Addition au memoire "Sur la convergence des series formees avec les iterees successives d'une fraction rationnelle" (Acta. tome 56) *Acta Math* 58 (1932). pp. 407 – 412.

[J 23] M.V. JAKOBSON. Transformation du cercle en luimeme. *Math Sbornik.* t. 85. no 2. 1971.

[J 24] E.I. JURY. *Theory and application of the Z-transform method.* Wiley. New York. 1964.

[J 25] E.I. JURY. B.W. LEE On the stability of a certain class on nonlinear sampled-data systems. *IEEE Trans. on Au. Control* 9 (1). 1964. pp. 51 – 61.

[J 26] E.I. JURY. *Inners and stability of dynamic systems. Wiley.* New York. 1974.

[K 1] R.E. KALMAN. J.E. BERTRAM. Control systems analysis and design via the 2nd method of Liapunov. Discretetime systems. *Transactions of the ASME* ser. D. v. 82. 1960. no 2.

[K 2] H. KAWAKAMI. Bifurcations of periodic solutions for 2nd order differential equations and its applications to parametric oscillations. *Congres ICNO. Berlin 1975.*

[K 8] G. KOENIGS. Recherches sur les substitutions uniformes. *Bull. Sci. Math.* (2) 7 (1883). pp. 340 – 358.

[K 9] G. KOENIGS. Sur une generalisation du theoreme de Fermat et ses rapports avec la theorie des substitutions uniformes. *Bull. Sci. Math.* (2) 8 (1884). pp. 286 – 288.

[K 10] G. KOENIGS. Sur les integrales de certaines equations fonctionnelles. *C.R. Acad. Sci. Paris.* 99 (1884). pp. 1016 – 1017.

[K 11] G. KOENIGS. Recherches sur les integrales de certaines equations fonctionnelles. *Am. Sci. Ecole Norm. Sup.* (3) 1 (1884). Supplement. pp. 341.

[K 12] G. KOENIGS. Nouvelles recherches sur les equations fonctionnelles. *Am. Sci. Ecole Norm. Sup.* (3) 2 (1885). pp. 285 – 404.

[K 13] G. KOENIGS. Sur les conditions d'holomorphisme des integrales de l'equation iterative. et de quelques autres equations fonctionnelles. *C.R. Acad. Sci. Paris.* 101 (1885). pp. 1137 – 1140.

[K 14] A.N. KOLMOGOROV. Conservation des mouvements periodiques conditionnels pour une petite variation de l'Hamiltonien. *Dokl. Akad. Nauk. SSSR.* t. 98. no 4. 1954.

[K 15] A.N. KOLMOGOROV. Theorie generale des systemes dynamiques et mecanique classique. *Congres International de Mathematiques. Amsterdam 1961.*

[K 17] M. KUCZMA. On the Schröder equation. *Rozprawy Mat.* 34 (1963). 50 pp.

[K 18] M. KUCZMA. Note on Schröder functional equation. *J. Australian Math. Soc.* 4 (1964). pp. 149 – 151.

[K 19] M. KUCZMA. Sur une equation aux differences finies et une caracterisation fonctionnelle des polynomes. *Fund. Math.* 55 (1964).. pp. 77 – 86.

[K 20] M. KUCZMA. *Functional equations in a single variable.* Polish Sc. Publ. Varsovie. 1968.

[K 21] M. KUCZMA. Divers aspects de l'equation fonctionnelle de Schröder. *Transformations Ponctuelles et Applications. (Colloque CNRS. Sept. 1973).* Editions CNRS Paris. 1976.

[K 22] M. KUCZMA. A. SMADJOR. Note on iteration of concave functions. *Amer. Math. Monthly.* 74 (1967). pp. 401 – 402.

[K 23] V.M. KUNSTEVITCH. Iu.N. TCHEHOVOI. Systemes non lineaires de commande a modulation d'impulsions en frequence et en largeur. *Technika.* Kiev. 1970.

[L 3] S. LATTES. Sur une classe d'equations fonctionnelles. *C.R. Acad. Sci. Paris.* 137 (1903). pp. 905 – 908.

[L 4] S. LATTES. Sur les substitutions a trois variables et les courbes invariantes par une transformation de contact. *Ibid.* 140 (1905). pp. 29 – 32.

[L 5] S. LATTES. Sur les equations fonctionnelles qui definissent une courbe ou une surface invariante par une transformation. *Ann. di Matematica.* (3) 13 (1906). pp. 11 – 37.

[L 6] S. LATTES. Sur la convergence des relations de recurrence. *C.R. Acad. Sci. Paris.* 150 (1910). pp. 1106 – 1109.

[L 7] S. LATTES. Sur les series de Taylor a coefficients recurrents. *C.R. Acad. Sci. Paris.* 150 (1910). pp. 1413 – 1415.

[L 8] S. LATTES. Sur les formes reduites des transformations ponctuelles a deux variables. Application a une classe remarquable de series de Taylor. *C.R. Acad. Sci. Paris.* 152 (1911). pp. 1566 – 1569.

[L 9] S. LATTES. Sur les formes reduites des transformations ponctuelles dans le domaine d'un point double. *Bull. Soc. Math. France.* 39 (1911). pp. 309 – 345.

[L 10] S. LATTES. Sur les suites recurrentes non lineaires et sur les fonctions generatrices de ces suites. *Ann. Fac. Sci. Toulouse.* (3) 3 (1912). pp. 73 – 124.

[L 11] S. LATTES. Sur l'iteration de substitutions rationnelles et les fonctions de Poincare. *C.R. Acad. Sci. Paris.* 166 (1918). pp. 26 – 28.

[L 12] S. LATTES. Sur l'iteration des substitutions rationnelles a deux variables. *C.R. Acad. Sci. Paris.* 166 (1918). pp. 151 – 153.

[L 13] S. LATTES. Sur l'iteration des fractions irrationnelles. *C.R. Acad. Sci. Paris.* 166 (1918). pp. 486 – 489.

[L 14] L. LEAU. Sur les equations fonctionnelles. *C.R. Acad. Sci. Paris.* 120 (1895). pp. 427 – 429.

[L 15] L. LEAU. Etude sur les equations fonctionnelles a une ou plusieurs variables. *Am. Fac. Sci. Toulouse.* 11 (1897). pp. 1 – 110.

[L 16] L. LEAU. Sur un probleme d'iteration. *Bull. Soc. Math. France.* 26 (1898). pp. 5 – 9.

[L 17] E.M. LEMERAY. Un theoreme sur les fonctions iteratives. *Bull. Soc. Math. France.* 23 (1895). pp. 255 – 262.

[L 18] E.M. LEMERAY. Sur les fonctions iteratives et sur une nouvelle fonction. *Assoc. Francaise pour l'Avancement des Sci. C.R. 24e session (Bordeaux 1895)* 24 (1896). pp. 149 – 165.

[L 19] E.M. LEMERAY. Sur la convergence des substitutions uniformes. *C.R. Acad. Sci. Paris.* 124 (1897). pp. 1220 – 1222.f

[L 20] E.M. LEMERAY. Sur la convergence des substitutions uniformes. *Nouv. Ann. de Math.,* (3) 16 (1897). pp. 306 – 319.

[L 21] E.M. LEMERAY. Sur quelques algorithmes et sur l'iteration. *Bull. Soc. Math. France.* 26 (1898). pp. 10 – 15.

[L 22] E.M. LEMERAY. Sur la convergence des substitutions uniformes. *Nouv. Ann. de Math.* (3) 17 (1898). pp. 75 – 80.

[L 23] E.M. LEMERAY. Sur quelques algorithmes generaux et sur l'iteration. *C.R. Acad. Sci. Paris.* 126 (1898). pp. 510 – 512.

[L 24] E.M. LEMERAY. Probleme d'iteration. *C.R. Acad. Sci. Paris.* 128 (1899). pp. 278 – 279.

[L 25] E.M. LEMERAY. Application des fonctions doublement periodiques a la solution d'un probleme d'iteration. *C.R. Acad. Sci. Paris.* 27 (1899). pp. 282 – 285.

[L 29] T. LEVICIVITA. Sopra alcuni criteri di instabilita. *Annali de Matematica.* Ser. 3. t. 5. 1901. pp. 221 – 305.

[L 31] P. LEVY. Sur l'iteration de la fonction exponentielle. *C.R. Acad. Sci. Paris.* 184 (1927). pp. 663 – 665. Fonctions a croissance reguliere et iteration d'ordre fractionnaire. *Ann. Mat. Pura Appl.* (4) 5 (1928). pp. 269 – 298. Fonctions a croissance reguliere et iteration d'ordre fractionnaire. *Atti del Congresso Internazionale dei Matematici, Bologna. 310 Septembre 1928.* Bologna 1930. pp. 277 – 282.

[L 33] A.M. LIAPUNOV. Probleme general de la stabilite du mouvement. *Ann. de la Fac. des Sc. de Toulouse.* vol. 9. 1907.

[L 34] A.M. LIAPUNOV. *Etude d'un cas particulier du probleme de la stabilite du mouvement.* Editions Universitaires Leningrad 1963.

[M 2] J. MATHER. *Structural stability of mappings.* Princeton 1966.

[M 3] R.M. MAY. Biological populations with non overlapping generations. *Science*, vol. 126. 1974. pp. 645 – 648.

[M 4] R.M. MAY. Simple mathematical models with very complicated dynamics. *Nature*, vol. 261, June 10, 1976.

[M 5] R.M. MAY. *Theoretical ecology.* Blackwell Oxford 1976.

[M 6] R.M. MAY. Thresholds and breakpoints in ecosystems with a multiplicity of stable states. *Nature*, vol. 269, 6 Oct. 1977, pp. 471 – 477.

[M 12] V.K. MEL'NIKOV. Sur l'existence de trajectoire doublement asymptotiques. *Dokl. Akad. Nauk. SSSR.* t. 211. No 5. 1973.

[M 13] V.K. MEL'NIKOV. On the existence of doubleasymptotic solutions for the Hamiltonian system. *Transformations Ponctuelles et Applications. (Colloque CNRS Sept. 73. Toulouse).* Editions CNRS Paris 1976.

[M 14] N. METROPOLIS. M.L. STEIN. P.R. STEIN. On finite limit sets for transformations on the unit interval. *Journal of Combin. Th.* (A). 15. 1973. pp. 25 – 44.

[M 15] C. MIRA. Extension des notions de points singuliers aux equations aux differences. *C.R. Acad. Sc. Paris.* t. 256. p. 3809 – 3812. 1963.

[M 16] C. MIRA. Condition de decroissance des solutions d'un systeme d'equations aux differences nonlineaires. *C.R. Acad. Sci. Paris.* t. 258. pp. 410 – 411. Groupe 1. 1963.

[M 17] C. MIRA. Determination pratique du domaine de stabilite d'un point d'equilibre d'une recurrence nonlineaire du deuxieme ordre a variables reelles. *C.R. Acad. Sci. Paris.* t. 261. p. 53145317: Groupe 2. 1964.

[M 18] C. MIRA. Sur quelques proprietes de la frontiere du domaine de stabilite d'un point double d'une recurrence et sur un cas de bifurcation de cette frontiere. *C.R. Acad. Sci. Paris.* t. 261. p. 53145317: Groupe 2. 1964.

[M 19] C. MIRA. F. ROUBELLAT. Etude d'un cas critique pour des recurrences du deuxieme ordre. *C.R. Acad. Sci. Paris.* t. 266. 1968. Serie A. p. 302 – 305.

[M 20] C. MIRA. Etude de la frontiere de stabilite d'un point double stable d'une recurrence nonlineaire autonome du deuxieme ordre. *Communication au Congres "International Pulse Symposium," Budapest, 9 – 11 Avril 1968.* D. 437/11. pp. 1 – 28.

[M 21] C. MIRA. Complex dynamics in two-dimensional endomorphisms. *Non Linear Analysis Theory, Methods & Applications.* Vol. 4(6), 1980, pp. 1167 – 1187.

[M 22] C. MIRA. F. ROUBELLAT. Cas ou le domaine de stabilite d'un ensemble limite attractif d'une recurrence du deuxieme ordre n'est pas simplement connexe. *C.R. Acad. Sci. Paris.* t. 268. pp. 1657 – 1660. Serie A. 1969.

[M 23] C. MIRA. Frontiere floue separant les domaines d'attraction de deux attracteurs. Exemples. *C.R. Acad. Sci. Paris.* t. 288. Serie A. 1979. pp. 591 – 594.

[M 24] C. MIRA. Etude d'un premier cas d'exception pour une recurrence, ou transformation ponctuelle, du deuxieme ordre. *C.R. Acad. Sci. Paris.* t. 269. pp. 1006 – 1109. Serie A. 1969.

[M 25] C. MIRA. Etude d'un second cas d'exception pour une recurrence, ou transformation ponctuelle, du deuxieme ordre. *C.R. Acad. Sci. Paris.* t. 270. pp. 332 – 335. Serie A. 1970.

[M 26] C. MIRA. Sur les cas d'exception d'une recurrence, ou transformation ponctuelle, du deuxieme ordre. *C.R. Acad. Sci. Paris.* t. 270. pp. 466 – 469. Serie A. 1970.

[M 27] C. MIRA. F. ROUBELLAT. Etude de la traversee d'un cas critique pour une recurrence du deuxieme ordre, sous l'effet d'une variation de parametre. *C.R. Acad. Sci. Paris.* t. 267. pp. 969 – 972. Serie A. 1968.

[M 28] C. MIRA. Traversee d'un cas critique, pour une recurrence du deuxieme ordre, sous l'effet d'une variation de parametre. *C.R. Acad. Sci. Paris.* t. 268. pp. 621 – 624. Serie A. 1969.

[M 29] C. MIRA. C. BURGAT. Sur une methode de determination des exposants caracteristiques d'un systeme d'equations differentielles lineaires a coefficients periodiques. *C.R. Acad. Sci. Paris.* t. 271. Serie A. pp. 965 – 968. 9 Novembre 1970.

[M 30] C. MIRA. L. PUN. Sur un cas critique d'une recurrence d'ordre m.m 2. possedant deux multiplicateurs complexes conjugues de module unite. *C.R. Acad. Sci. Paris.* t. 272. Serie A. pp. 505 – 508. 15 Fevrier 1971.

[M 31] C. MIRA. Cas critique d'une recurrence ou d'une transformation ponctuelle, du quatrieme ordre avec multiplicateurs complexes. *C.R. Acad. Sci. Paris*. t. 272. Serie A. pp. 1727 – 1730. 28 Juin 1971.

[M 32] C. MIRA. Sur un cas critique correspondant a deux multiplicateurs egaux a l'unite pour une recurrence ou transformation ponctuelle de deuxieme ordre. *C.R. Acad. Sci. Paris*. t. 275. Serie A. pp. 85 – 88. 3 Juillet 1972.

[M 33] C. MIRA. Sur les courbes invariantes fermees des recurrences nonlineaires, voisines d'une recurrence lineaire conservative du 2e ordre. *Transformations Ponctuelles et Applications. (Colloque CNRS Sept. 73, Toulouse)*. Editions CNRS Paris. 1976.

[M 34] C. MIRA. Structure de bifurcations "boitesemboitees" dans les recurrences du premier ordre dont la fonction presente un seul extremum. *C.R. Acad. Sci. Paris*. t. 282. Serie A (Jan. 1976). pp. 219 – 222.

[M 35] C. MIRA. Etude d'un modele de croissance d'une population biologique. *C.R. Acad. Sci. Paris*. t. 282. Serie A (1976). pp. 1441 – 1444.

[M 36] C. MIRA. Accumulations de bifurcations et structures boitesemboitees dans les recurrences et transformations ponctuelles. *Proceedings of the VIIe International Conference on Nonlinear Oscillation. Berlin. Sept. 1975* (Akademic Verlag. Berlin 1977).

[M 37] C. MIRA. Sur la double interpretation, deterministe et statistique, de certaines bifurcations complexes. *C.R. Acad. Sci. Paris*. t. 283. Serie A. 1976. pp. 911 – 914.

[M 38] C. MIRA. Sur la notion de frontiere floue de stabilite. *Proceedings of the third Brazilian Congress of Mechanical Engineering. Rio de Janeiro. Dec. 1975*. D4. pp. 905 – 918.

[M 39] C. MIRA. Systemes a dynamique complexe et bifurcations boitesemboitees. *Iere partie RAIRO Automatique. Mars 1978. vol. 12. no 12e partie. Juin 1978*. vol. 12. no 2.

[M 40] C. MIRA. Dynamique complexe engendree par une equation differentielle d'ordre 3. *Actes du Congres Equadiff 78, Florence. Mai 1978*.

[M 41] C. MIRA. Bifurcations boitesemboitees dans les endomorphismes definis par un polynome du 3e degre. *Note interne no 2 du Groupe Systemes Dynamiques NonLineaires et Applications. Avril 1977.* Universite PaulSabatier, Toulouse.

[M 42] C. MIRA. *Cours de Systemes Asservis NonLineaires.* Dunod Universite. 1969.

[M 45] P. MONTEL. L'iteration. Univ. Nac. La Plata. *Publ. Fac. Ci. Fisicomat Revista.* (2) 3, 1948, pp. 201 – 211.

[M 46] P. MONTEL. *Lecons sur les recurrences et applications.* GauthierVillars. 1957. Paris.

[M 47] J. MOSER. The analytic invariants of an area-preserving mapping near a hyperbolic fixed point. *Comm. Pure Appl. Math.* 9 (1956). pp. 673 – 692.

[M 48] J. MOSER. On invariant curves of areapreserving mappings of an annulus. *Nachr. Akad. Wiss. Gottingen, Math. Phys.* K1. II 1962. pp. 1 – 20.

[M 49] J. MOSER. *Stable and Random Motion in Dynamical Systems with special emphasis on Celestial Mechanics.* Princeton University Press, 1973.

[M 50] P.J. MYRBERG. Iteration der reellen Polynome zweiten Grades I.II.III. *Ann. Acad. Sci. Fenn.* Ser A. 256 (1958), pp. 1 – 10; 268 (1959), pp. 1 – 10; 336 (1963), pp. 1 – 10.

[M 51] P.J. MYRBERG. Iteration von Quadratwurzeloperationen. *Ann. Acad. Sci. Fenn.* Ser A. 259 (1958), pp. 1 – 10.

[M 52] P.J. MYRBERG. Inversion der Iteration fur rationale Funktionen. *Ann. Acad. Sci. Fenn.* Ser A. 292 (1960), pp. 1 – 14.

[M 53] P.J. MYRBERG. Sur l'iteration des polynomes reels quadratiques. *J. Math. Pures Appl..* (9), 41 (1962), pp. 339 – 351.

[M 54] P.J. MYRBERG. Iteration der Binome beliebigen Grades. *Ann. Acad. Sci. Fenn.* Ser A.I., 348 (1964), pp. 1 – 14.

[M 55] P.J. MYRBERG. Iteration der Polynome mit reellen koeffizienten. *Ann. Acad. Sci. Fenn.* Ser A.I., 374 (1965), pp. 1 – 18.

[M 57] C. MIRA. Recurrences et bifurcations d'un domaine d'attraction avec modifications qualitatives importantes. *Colloque sur les comportements generaux des processus iteratifs. La Garde Freinet. 21 – 23 Mai 1979.*

[M 58] C. MIRA. Dynamique complexe engendree par une recurrence, continue, lineaire par morceaux du premier ordre. *C.R. Acad. Sc. Paris*. t. 285. Serie A. 1977. pp. 731 – 734.

[M 59] C. MIRA. Sur la structure des bifurcations des diffeomorphismes du cercle. *C.R. Acad. Sc. Paris*. t. 287. Serie A. 1978. pp. 883 – 886.

[M 60] C. MIRA. Sur quelques problemes de dynamique complexe. *Colloque "Modeles mathematiques en biologie". Journees Math. de la Societe Math. de France. Montpellier 2224 Nov. 1978*. Lecture Notes in Biomathematics 41, pp. 169-205, Springer 1980.

[M 61] C. MIRA. Dynamique complexe dans les endomorphismes a 2 dimensions. *ICNO 78 Prague, 1115 Sept. 1978*.

[N 1] Yu.I. NEIMARK. *Methode des transformations ponctuelles dans la theorie des oscillations non lineaires*. Nauka. Moscou. 1972.

[N 2] Yu.I. NEIMARK. I.M. KUBLANOV. Regimes periodiques, stabilite, pour un systeme a parametres repartis a relais destine a reguler la temperature. *Avtom.i Tel.*, t. 14. no 1. 1953.

[N 3] Yu.I. NEIMARK. M.I. FEIGIN. Sur un cas de bifurcation dans les systemes a relais. *Radiofisica*. t. 7. no 2. 1964.

[N 4] Yu.I. NEIMARK. Methode des transformations ponctuelles dans la theorie des oscillations non lineaires. Parties I.II.III. *Radiofisica*. t. 1. no 1; t. 2. no 2; t. 1. no 56, 1958.

[N 5] Yu.I. NEIMARK. Methode des transformations ponctuelles dans la theorie des oscillations nonlineaires. *Actes de International Conference On NonLinear oscillations (ICNO)*. T. 2. Editions de l'Acad. des Sc. d'Ukraine. Kiev 1963.

[N 6] Yu.I. NEIMARK. Sur une classe de systemes dynamiques. *Actes de l'ICNO*. t. 2: editions de l'Acad. des Sc. d'Ukraine. Kiev 1971.

[N 7] Yu.I. NEIMARK. Etude de la stabilite d'un point fixe d'une transformation dans un cas critique. *Radiofisica*. t. 3. no 2. 1960.

[N 8] Yu.I. NEIMARK. Sur un type de bifurcation dans les systemes a relaix. *Radiofisica*. t. 7. no 2. 1964.

[N 9] Yu.I. NEIMARK. S.D. KINJAPIN. Sur un etat d'equilibre situe sur une surface de discontinuite. *Radiofisica*. t. 3. no 4. 1960.

[N 10] Yu.I. NEIMARK. S.D. KINJAPIN. Naissance d'un regime periodique a partir d'un etat d'equilibre situe sur une surface de discontinuite. *Radiofisica*. t. 5. no 6. 1962.

[N 11] Yu.I. NEIMARK. Methode de la moyenne du point de vue des transformations ponctuelles. *Radiofisica*. t. 6. no 5. 1963.

[N 12] Yu.I. NEIMARK. Quelques cas de dependance d'un regime periodique par rapport a des parametres. *Dokl. Akad. Nauk SSSR*. t. 129. no 4. 1959.

[N 13] Yu.I. NEIMARK. Lien entre de petites perturbations d'un systeme d'equations differentielles, et la transformation ponctuelle correspondante. *Dokl. Akad. Nauk SSSR*. t. 148. no 2. 1963.

[N 14] Yu.I. NEIMARK. L.P. SHILNIKOV. Etude des systemes dynamiques voisins du cas lineaire par morceaux. *Radiofisica*. t. 3. no 4. 1960.

[N 15] Yu.I. NEIMARK. L.P. SHILNIKOV. Application de la methode du petit parametre a des equations differentielles avec second membre discontinu. *Izv. Akad. Nauk SSSR OTN* no 6. 1959.

[N 16] Yu.I. NEIMARK. Existence et grossierete des varietes invariantes des transformations ponctuelles. *Radiofisica*. t. 10. no 3. 1967.

[N 17] Yu.I. NEIMARK. Variete integrale des equations differentielles. *Radiofisica*. t. 10. no 3. 1967.

[N 18] Yu.I. NEIMARK. Mouvements proches d'un mouvement doublement asymptotique. *Dokl. Akad. Nauk. SSSR*. t. 172. no 5. 1967.

[N 19] T. NISHINO. T. YOSHIOKA. Sur l'iteration des transformations rationnelles de l'espace de deux variables complexes. *Ann. Scient. Ec. Norm. Sup.* 3e ser. t. 82. 1965. pp. 327 – 376.

[N 20] N.E. NORLUND. *Lecons sur les equations lincaires aux differences finies*. GauthierVillars. Paris. 1929.

[P 1] A.M. PANOV. Comportement des trajectoires des systemes d'equations aux differences finies au voisinage d'un point singulier. *Utch. Zap. Ural'skovo Gos. Univ.* 19. 1956.

[P 2] A.M. PANOV. Classification des points singuliers des equations aux differences finies a n dimensions. *Utch. Zap. Ural'skovo Gos. Univ.* 23. 1959. pp. 13 – 21.

[P 3] A.M. PANOV. Comportement de la solution d'un systeme d'equations aux differences finies au voisinage d'un point fixe. *Matematika.* 1959. no 5. pp. 173 – 183.

[P 4] A.M. PANOV. Etude qualitative des trajectoires des equations aux differences finies au voisinage d'un point fixe. *Matematika.* 1960. no 1. pp. 166 – 174.

[P 5] A.M. PANOV. Sur une classe de systemes iteratifs. *Matematika.* 1962. no 3. pp. 111 – 115.

[P 6] A.M. PANOV. Comportement qualitatif des trajectoires d'un systeme d'equations aux differences au voisinage d'un point singulier. *Matematika.* 1964. no 3. pp. 111 – 115.

[P 7] A.M. PANOV. Sur l'indice des systemes iteratifs. *Matematika.* 1965. no 2. p. 76.

[P 9] H. POINCARE. Sur les courbes definies par des equations differentielles. *Journ. de Math. pures et appl.*, t. 1. 1885. pp. 167 – 244.

[P 10] H. POINCARE. *Les methodes nouvelles de la mecanique celeste.* T. 1.2.3. Paris 1892. 1893. 1899.

[P 14] C.P. PULKIN. Suites iteratives oscillantes. *Dokl. Akad. Nauk SSSR.* t. 23. no 6. 1950.

[R 4] O.E. RÖSSLER. Chaotic behaviour in simple reaction systems. *Z. Naturforsch.* 31 a. 1976. pp. 256 – 264.

[R 5] O.E. RÖSSLER. An equation for continuous chaos. *Phys. Letters.* vol. 57A. no 5. 1976. pp. 397 – 398.

[R 6] O.E. RÖSSLER. Chemical turbulence: chaos in a simple reactiondiffusion system. *Z. Naturforsch.* 31 a. 1976. pp. 1168 – 1172.

[R 7] O.E. RÖSSLER. Chaos in the Zabotinski reaction. *Nature.* vol. 271. 5 Jan. 1978. p. 89.

[R 12] O.E. RÖSSLER. Chaos in abstract kinetics. Two prototypes. *J. Math. Biol.*, vol. 39. 1977. pp. 275 – 289.

[R 13] O.E. RÖSSLER. Continuous chaos, in "Bifurcation theory and applications in scientific disciplines." *Ann. of the N.Y. Acad. of Sc.*, vol. 316. 1979. pp. 376 – 392.

[R 14] O.E. RÖSSLER. An equation for hyperchaos. *Physics Letters*, vol. 71A. no 2.3. 1979. pp. 155 – 157.

[R 15] O.E. RÖSSLER. Different kinds of chaotic oscillations in the BelousovZabotinskii Reaction. *Z. Naturforsch,* 33a. 1179 – 1183. 1978.

[R 16] O.E. RÖSSLER. Chaos, in *Structural Stability in Physics.* Springer 1979.

[S 2] P.A. SAMUELSON. *Foundations of economic analysis*, Harvard University Press. 1947.

[S 3] E. SCHRÖDER. Uber iterierte functionen. *Math. Ann.* 2. 1871. pp. 296 – 322.

[S 4] A.N. SHARKOVSKII. Condition necessaire et suffisante de convergence d'une iteration a une dimension. *Ukrainskii Mat. Jurnal.* t 12. no 4. 1960. pp. 484 – 489.

[S 5] A.N. SHARKOVSKII. Reductibilite d'une fonction continue de la variable reelle et structure des points fixes du processus iteratif associe. *Dokl. Akad. Nauk SSSR.* t. 139. no 5. 1961. pp. 1067 – 1070.

[S 6] A.N. SHARKOVSKII. Systeme partiellement ordonne d'ensembles attractifs. *Dokl. Akad. Nauk SSSR.* t. 170. no 6. 1966.

[S 7] A.N. SHARKOVSKII. Processes iteratifs a convergence rapide. *Ukrainskii Mat. Jurnal.* t. 13. no 2. 1961. pp. 210 – 215.

[S 8] A.N. SHARKOVSKII. Coexistence des cycles d'une transformation continue d'une droite, en elle-meme. *Ukrainskii Mat. Jurnal.* t. 16. no 1. 1964. pp. 61 – 71.

[S 9] A.N. SHARKOVSKII. Cycles et structure d'une transformation continue. *Ukrainskii Mat. Jurnal.* t. 17. no 3. 1965. pp. 104 – 111.

[S 10] A.N. SHARKOVSKII. Classification des points fixes. *Ukrainskii Mat. Jurnal.* t. 17. no 5. 1965. pp. 80 – 95.

[S 11] A.N. SHARKOVSKII. Comportement d'une transformation au voisinage d'un ensemble attractif. *Ukrainskii Mat. Jurnal.* t. 18. no 2. 1966. pp. 60 – 82.

[S 12] A.N. SHARKOVSKII. A propos d'un theoreme de Birkhoff. *Dop. Akad. Nauk Ukr.* R.S.R. Kiev. 1967. pp. 429 – 432.

[S 13] A.N. SHARKOVSKII. Ensemble attractif ne contenant pas de cycle. *Ukrainskii Mat. Jurnal.* t. 20. no 1. 1968. pp. 136 – 142.

[S 14] L.P. SHILNIKOV. Cas de naissance de solutions periodiques a partir d'une trajectoire singuliere. *Mat. Sbornik.* t. 61. 104. 1963. pp. 443 – 466.

[S 15] L.P. SHILNIKOV. Yu.I. NEIMARK. Cas de naissance de regimes periodiques. *Dokl. Akad. Nauk SSSR.* t. 160. no 6. 1965. pp. 1261 – 1264.

[S 16] L.P. SHILNIKOV. Yu.I. NEIMARK. Cas de naissance de regimes periodiques. *Radiofisika.* t. 8. no 2. 1965. pp. 330 – 340.

[S 17] L.P. SHILNIKOV. Sur un cas d'existence d'un ensemble denombrable de regimes periodiques. *Dokl. Akad. Nauk SSSR.* t. 160. no 3. 1965. pp. 558 – 561.

[S 18] L.P. SHILNIKOV. Naissance d'un mouvement periodique a partir de la trajectoire qui va d'un etat d'equilibre colcol et y revient. *Dokl. Akad. Nauk SSSR.* t. 170. no 1. 1966. pp. 49 – 52.

[S 19] L.P. SHILNIKOV. Sur un probleme de PoincareBirkhoff. *Mat. Sbornik.* t. 74 (116). no 3. 1967. pp. 378 – 397.

[S 20] L.P. SHILNIKOV. Existence d'un ensemble denombrable de regimes periodiques dans l'espace a quatre dimensions, au voisinage etendu d'un colfoyer. *Dokl. Akad. Nauk SSSR.* t. 172. no 1. 1967. pp. 54 – 57.

[S 21] L.P. SHILNIKOV. Existence d'un ensemble denombrable de regimes periodiques au voisinage d'une courbe homocline. *Dokl. Akad. Nauk SSSR.* t. 172. no 2. 1967. pp. 298 – 301.

[S 22] L.P. SHILNIKOV. Probleme de la structure du voisinage d'un tube homocline d'un tore invariant. *Dokl. Akad. Nauk SSSR.* t. 180. no 2. 1968. pp. 286 – 289.

[S 23] L.P. SHILNIKOV. Naissance d'un mouvement periodique a partir d'une trajectoire doublement asymptotique a un etat d'equilibre de type col. *Mat. Sbornik.* t. 77 (119). no 3. 1968. pp. 461 – 472.

[S 24] L.P. SHILNIKOV. Nouveau type de bifurcation des systemes dynamiques multidimensionnels. *Dokl. Akad. Nauk SSSR.* t. 189. no 1. 1969. pp. 59 – 692.

[S 25] L.P. SHILNIKOV. Probleme de la structure du voisinage etendu d'un etat d'equilibre grossier de type col-foyer. *Mat. Sbornik*. t. 81 (123), no 1. 1970. pp. 92 – 103.

[S 26] L.P. SHILNIKOV. N.K. GAVRILOV. Systemes dynamiques tridimensionnels, voisins de systemes avec courbe homocline non grossiere. *Mat. Sbornik*. t. 88 (130). no 4. 1972. pp. 476 – 492 et t. 90 (132). no 1. 1973. pp. 139 – 156.

[S 27] L.P. SHILNIKOV. Theorie des bifurcations des systemes dynamiques et frontieres dangereuses. *Dokl. Akad. Nauk SSSR*. t. 224. no 5. 1975. pp. 1046 – 1049.

[S 30] S. SMALE. Morse inequalities for a dynamical system. *Bull. A.M.S.* t. 66. 1960. pp. 43 – 49.

[S 31] S. SMALE. Differentiable dynamical systems. *Bull. A.M.S.* t. 73. 1967. pp. 747817.

[S 33] P.R. STEIN. S. ULAM. Lectures in nonlinear algebraic transformations. *Lectures Nato Adv. Studies Institutes*, Istambul. 1970. D. Reitel Publ. Co. (1973).

[T 7] Yu.N. TCHEKOVOI. Asymptotic stability in the large of pulse-frequency modulated system of the 2nd type. *Transformations Ponctuelles et Applications. (Colloque CNRS, Sept. 73)*. Editions CNRS. Paris 1976.

[T 9] Ya.z. TSYPKIN. Theorie des systemes a impulsions. *Fizmatgiz*. Moscou 1958.

[T 10] Ya.z. TSYPKIN. *Methode de Goldfarb et application a l'analyse des regimes periodiques dans les systemes a impulsions*. G.E.I. 1962.

[U 1] B. ULEWICZ. On conditions for attractive singular points. *Transformations Ponctuelles et Applications. (Colloque CNRS, Sept. 73, Toulouse)*. Editions CNRS. Paris 1976.

[U 2] M. URABE. Periodic solutions to a certain difference equation and their applications to pseudoperiodic differential equations. *Transformations Ponctuelles et Applications (Colloque CNRS, Sept. 73, Toulouse)*. Editions CNRS. Paris 1976.

[V 1] P. VIDAL. *Nonlinear sampleddata systems*. Gordon and Breach 1969.

[Z 2] M.C. ZDUN. On the orbits of hatfunctions. *Colloquium Math.*, vol. 36. fasc. 2. 1976. pp. 250 – 254.

HISTORY, PART 2

SOME HISTORICAL ASPECTS CONCERNING THE THEORY OF DYNAMIC SYSTEMS[1]

A6.1 INTRODUCTION

Dynamics is a concise term referring to the study of time-evolving processes, and the corresponding system of equations, which describes this evolution, is called a *dynamic system*. Initially, the theory of dynamic systems builds on the foundations of the results of Poincaré (1878–1900), Liapunov (1893), Birkhoff (1908–1944) and those concerning the maps, iterations, and recurrences obtained at the end of the 19th century and the beginning of the 20th century by Koenigs, Lemeray, Lattes, and Hadamard. Since then it has had its most spectacular and organized growth within the framework of two Soviet schools of thought (from 1920): the Mandelstham-Andronov school (Moscow-Gorki), and the Krylov-Bogoliubov school (Kiev). Very strong interactions between the theoretical researches and practical implications in the physical, or engineering systems, are the reason for the success of these two schools. They have led the Poincaré-Liapunov-Birkhoff methods to the highest development in what they called the *theory of nonlinear oscillations*, the most important component of the *theory of dynamic systems*. In dynamics, these two schools occupy incontestably the

1. This is another historical essay by Christian Mira, written in 1986, and giving further detail on the Russian contributions. It appeared in *Dynamical Systems: A Renewal of Mechanism Centennial of George David Birkhoff*, S. Diner, D. Fargue, and G. Lochak, eds. Singapore: World Scientific.

first place, which is admitted by the most famous American mathematicians (for example, cf. J.P. Lasalle and S. Lefschetz, J. of Math. Anal. and Appl., 2 (1961), pp. 467 – 499).

From 1960, and specially since 1975 with the explosive growth of research in the *chaotic dynamic* field, and with the translation of the Soviet results in the Western countries, the study of dynamic systems has become a subject in vogue outside of USSR, with more and more papers concerning all the scientific disciplines. Nevertheless, in spite of a practical motivation often announced by the authors, most of these papers concerns "*abstract dynamic systems*," and are devoted to nonessential generalizations without any interest for understanding a typical dynamic behavior, or for practical purposes. This is one of the characteristics of the present period, as opposed to the study of "*concrete dynamic systems*," which motivated the Mandelstham-Andronov-Krylov-Bogoliubov schools.

This appendix is devoted to some historical aspects related to the evolution of the dynamic systems theory. The first part concerns the contribution of G.D. Birkhoff. The second deals with the above-mentioned Soviet schools. As a third step, in connection with the sudden interest since 1975 in one-dimensional endomorphisms defined by an unimodal function, Myrberg's fundamental contribution (1958 – 1963) is recalled. In particular, Myrberg's work is devoted to the phenomena of period-doubling cascades, which represents, through the Poincaré method of surface sections, a typical behavior related to continuous systems dominated by strong nonlinearities. It is surprising that Myrberg's results are systematically passed over in silence in the contemporary papers dealing with this subject in vogue since 1978. So, while reviewing the historical aspects of the subject, this appendix intends to try to correct certain bad habits of citation adopted in contemporary publications.

In this appendix, we consider that the field of "*concrete dynamic systems*" consists of two parts. The first is related to the study of problems directly suggested by applications (physics, engineering, etc.). The second part concerns the theoretical study of equations, having the lowest dimension and the simplest structure, simplified by eliminating the more complicated structure of real-world examples. The phenomenon of period-doubling cascades, and their accumulations with the very simple discrete dynamic system $x_{n+1} = x_n^2 - \lambda$ (x being a real variable and (λ a real parameter, $n = 0, 1, 2,...$) is an example of the benefit of this approach, and it has many practical implications.

A6.2 G.D. BIRKHOFF

Researches about the theory of dynamic systems represent an important part of Birkhoff's contribution to the mathematical sciences. Though not directly related to practice, these researches can be considered in the field of "concrete dynamic systems" for the above-mentioned reasons regarding nonessential generalizations. Indeed, for Birkhoff, the systematic organization or presentation of a theory was secondary in importance to its discovery, which is not the case nowadays.

Intellectually, Birkhoff was the greatest disciple of Poincaré, taking over the techniques and problems of Poincaré and carrying on that work. So, he began with the very nice proof (1913) of Poincaré's Last Theorem (1912) concerning the restricted problem of three bodies, which was proved by Poincaré only for very special cases. This proof and the *Ergodic Theorem* (1931) constitute the most popular results of Birkhoff. Nevertheless, his most important contribution in dynamics concerns his extension of Poincaré's work in celestial mechanics, his formal theory of stability, and the consequent geometrical theory, his results about difference equations.

So, among the greatest advances in this field are Birkhoff's results on the fixed points, or periodic orbits, of *saddle* (hyperbolic) type, or *center* (elliptic) type, the *homoclinic* and *heteroclinic* structures, the *unstable centers*, the *instability rings*, the concept of *signature* associated with a homoclinic structure, for the two-dimensional surface transformations. This also true for his concepts of *minimal*, or *recurrent sets*, and *wandering*, *central*, or *transitive motions*.

In his book *Dynamical Systems* (1927), Birkhoff wrote:

> *the final aim of the theory of motions of a dynamical system must be directed towards the qualitative determination of all possible types of motions and the interrelation of these motions.*

The definition and classification of all possible types of dynamic motions constitute an important contribution of this author. Using ideas developed by Poincaré, a classification of dynamic motions was proposed by Birkhoff in 1927, and refined by Andronov in 1933; this classification is represented in the diagram of Figure 1. In this diagram, an increase of the structural complex-

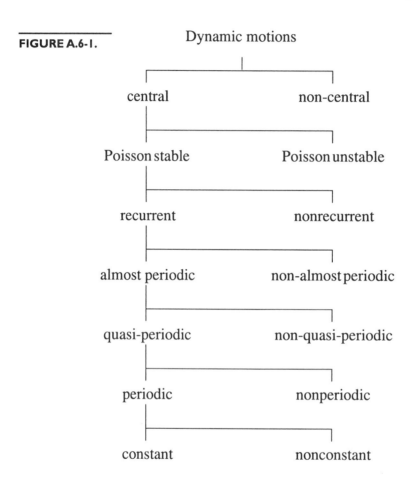

FIGURE A.6-1.

Dynamic motions

central — non-central

Poisson stable — Poisson unstable

recurrent — nonrecurrent

almost periodic — non-almost periodic

quasi-periodic — non-quasi-periodic

periodic — nonperiodic

constant — nonconstant

ity of the motions manifests itself as a gradual transition from orderly to chaotic (erratic, or stochastic) motions. It is conjectured that in general as the dimension of the dynamic system increases, the structural complexity of the motion increases. Nowadays, depending on the context, a motion is called chaotic when it is at least nonperiodic.

One can divide Birkhoff's research in dynamics into formal and nonformal aspects. The formal aspects may be regarded as relatively complete. His qualitative dynamical theory is restricted to the conservative case with two degrees of freedom ($m = 2$) due to the doubtful existence of a *regular surface of section* of Poincaré for more than two degrees of freedom ($m > 2$). But it seems very probable that such Poincaré surfaces exist in general for $m = 2$. This qualitative theory reaches its highest point with the prize paper, Nouvelles recherches sur les systemes dynamiques (*Memoriae Pont. Acad. Sci. Novi Lyncaei*, 1935, 53, vol. 1, pp. 85 – 216), which resumes and extends his earlier dynamical results. It is in this paper that he introduces the signature, a two-dimensional symbol which displays the topology of a homoclinic, or heteroclinic, structure. Such a symbol is the ultimate in the qualitative description of a dynamical system. He indicates also in this famous paper the formal analogies between the sets asymptotic to elliptic points (unstable centers) and the invariant curves passing through hyperbolic points. All the results about qualitative theory had implications for his research concerning *the restricted problem of three bodies*.

G.D. Birkhoff's Collected Mathematical Papers was published in three volumes by Dover in 1968, with a presentation of his mathematical work by Marston Morse (reprint from *Bull. Amer. Math. Soc.*, May 1946, vol. 52, no 5, part 1, pp. 357 – 391). This publication presents many new ideas, new theorems, and new questions of dynamics.

A6.3 NONLINEAR OSCILLATIONS FROM 1925

Except for Birkhoff's work in the United States, from 1925 the most important contribution to the theory of dynamic systems has been achieved in the USSR, mainly with the Mandelstham-Andronov school and the Krylov-Bogoliubov school. In connection with this, J.P. LaSalle and S. Lefschetz said in 1961:

In the USSR the study of differential equations has profound roots, and in this subject the USSR occupies incontestably the first place. One may also say that Soviet specialists, far from working in vacuum, are in intimate contact with applied mathematicians and front rank engineers. This has brought great benefits to the USSR and it is safe to say that the USSR has no desire to relinquish these advantages. (J. of Math. Anal. and Appl., 2, pp. 467-499, 1961.)

In the field of concrete dynamic systems, two different paths have been followed since 1925. The first one is classical, dating from the 19th century. It concerns the elaboration of particular methods fitted to the solution of specific problems in the physical or engineering sciences. The second line of research has no direct interest in the study of a particular phenomenon of a given scientific discipline. Its purpose is the search for analogies of behavior concerning the dynamic systems belonging to different branches of science, and the construction of general mathematical tools for describing and studying dynamical phenomena. This investigation gives rise to two different approaches.

The first approach corresponds to *qualitative methods*. To define the strategy of these methods, one has to note that the solutions of the equations of nonlinear dynamic systems are in general nonclassical transcendental functions which are very complex. The strategy is similar to that used to characterize a function of complex variables by its singularities (zeros, poles, essential singularities). Here, the complex transcendental functions are defined by the singularities of continuous (or discrete) dynamic systems, such as:

- stationary states which are equilibrium points (fixed points), or periodic solutions (cycles);

- trajectories (invariant curves) passing through saddle singularities of two-dimensional systems;

- stable and unstable manifold for dimensions greater than two;

- boundary, or separatrix, of the influence domain (domain of attraction, or basin) of a stable (attractive) stationary state;

- homoclinic or heteroclinic singularities;

- more complex singularities.

Qualitative methods consider the nature of these singularities in the phase space and their evolution when parameters of the system are varied.

The second approach corresponds to the *analytical methods*. Here, the above-mentioned complex transcendental functions are defined by convergent, or at least asymptotically convergent, or convergent in the mean, series expansions. The method small parameters of Poincaré, and the asymptotic methods of Krylov-Bogoliubov-Mitropolski are analytical; so are the averaging method and the method of harmonic linearization in the theory of nonlinear oscillations.

These two approaches constitute two relatively independent branches of the nonlinear oscillation theory. They are not purely mathematical, nor physical, but they have the same aims: the construction of mathematical tools for the solution of concrete problems, and the development of a general theory of dynamic systems.

Since 1960, the important development of computer methods has given a great expansion to the numerical approach to dynamic problems. Such an approach is a useful tool for the qualitative or analytical methods.

A6.4 THE MANDELSTHAM-ANDRONOV SCHOOL

A6.4.1   Generalities

The Mandelstham-Andronov school is a unique example of an efficient organization devoted to a long-term joint research project, involving both theory and applications. This evolved in the close collaboration of pure and applied mathematicians, top-echelon engineers, and physicists.

The problem of constructing mathematical tools for the study of nonlinear oscillations was firstly formulated by Mandelstham around 1920, in connection with the study of dynamic systems in radio-engineering. Indeed, with Papaleski, Mandelstham formulated the fundamental problems solved later by his disciples. Such formulations constituted a decisive step in the understanding of concrete dynamic systems.

In the beginning, the most popular approach to nonlinear problems in this school was the fitting method. This method is based on the approximation of a nonlinear characteristic by a piecewise linear characteristic. The solution of the nonlinear problem is reduced to the solution of a set of linear problems corresponding to the different segments, with conditions of continuity at the endpoints of the segments. With the help of this method, up until 1927 this school considered specific problems in the physical and engineering sciences.

In 1927, Andronov, the most famous of Mandelstham's students, presented his thesis on a topic formulated by Mandelstham, *Poincaré's limit cycles and the theory of oscillations*. This thesis is a contribution of the highest rank to the theory of nonlinear oscillations, because it opens a new path for applications of Poincaré's qualitative theory of differential equations, with many practical consequences. With this work Andronov was the first to see that self oscillations correspond to limit cycles.

Later Andronov amplified his activity with a precise purpose: *the elaboration of a theory of nonlinear oscillations*, in order to develop a mathematical tool, useful to different scientific disciplines. For this theory he used following foundations:

- Lyapunov's stability theory;

- Poincaré's qualitative theory of differential equations;

- map theory (the basic tool for Poincaré sections); and

- Birkhoff's classification of all possible types of dynamic motions.

A6.4.2 Contribution to the qualitative methods

Using ideas developed by Poincaré, Andronov showed, in the case of two-dimensional autonomous ordinary differential equations, that it is possible *to divide the phase plane into cells*, each filled with qualitatively identical trajectories. Each cell, closed by a curve boundary, is the basin of attraction of a stable steady state: an equilibrium point, or a limit cycle. In the physical sense, a cell corresponds to a given mode of operation of a concrete process. This result has no direct extension, however, to n-dimensional spaces, for n > 2, due to the difficulties of defining the boundary of the basins (which may be a fractal) in the general case.

The complete global study of the autonomous phase plane was made by Andronov, and later, Gordon and Mayer. The corresponding results, giving in particular the structure of trajectories in the critical case in Liapunov's sense (the linear approximation cannot be used to study an infinitely small neighborhood of a steady state) and the behavior on Poincaré's equator, are developed by these four authors in *Qualitative theory of dynamic systems* (Edor Nauka, Moscow, 1966).

One of the most spectacular contributions of the Mandelstham-Andronov school concerns the *theory of bifurcations*, part of the qualitative theory of dynamic systems.

Andronov, in 1937, was the first to study analytically the bifurcation giving rise to m limit cycles from a complex focus of multiplicity $k > m$ in the plane. This bifurcation was described by Poincaré (periodic solution of second type) and further studied by Hopf in 1942. The other typical bifurcations analyzed by Andronov, Leontovitch, Gordon, and Mayer were:

- birth of m limit cycles from a loop constituted by the separatrices of a saddle; or

- from a complex (multiple) limit cycle; or

- from a loop constituted by the separatrices of a saddle-node point; or

- from a closed trajectory of a conservative case.

Within the framework of the theory of bifurcations, A.A. Andronov and L.S. Pontriagin introduced in 1937 the concept of roughness or structural stability. This concept is essential for practice as well as for theory. Roughly speaking, a dynamic system is structurally stable if the topological structure of its motions does not change for small changes of the parameters, or the structure, of the equations describing these motions. Then, to be physically significant, a problem of a dynamic system must respect the following conditions:

- a solution should exist;

- this solution should be unique;

- the unique solution should be continuous with respect to the data contained in the initial conditions, or in the boundary conditions; and

- the dynamic system should be structurally stable.

The three first conditions were formulated by Hadamard in 1923.

The study of the problem of the structural stability can be considered complete for two-dimensional autonomous dynamic systems. In 1937 Andronov and Pontriagin formulated the corresponding theorems in the analytic case. In 1952, De Baggis gave the demonstration of these theorems in the more general case of continuous and differential functions defining the dynamic system for C^1-small perturbations.

General conditions of structural stability, for autonomous two-dimensional systems, are: the system has only a finite number of equilibrium points and limit cycles, which are not critical in the sense of Liapunov; no separatrix joins the same, or two distinct, saddle points. It is then possible to define, in the parameter space of the system, a set of cells inside which the qualitative behavior is preserved.

The knowledge of such cells is of great importance for the analysis and synthesis of dynamic systems in physics and engineering. On the boundary of a cell, the dynamic system is structurally unstable, and, for the autonomous two-dimensional systems (two-dimensional vector fields), the structurally stable systems are dense. Until 1966, the extension of this result to higher-dimensional systems was conjectured. But Smale's paper Structurally stable systems are not dense (*Amer. J. Math.* 88, 1966, 491– 496) showed that this conjecture is false in general. So it appears that, with an increase of the problem dimension, the complexity of the parameter space also increases. The boundaries of the cells defined in the phase space, as well as in the parameter space, have in general a complex (fractal) structure for *n*-dimensional vector fields, *n* > 2.

For autonomous two-dimensional dynamic systems, the results of the Mandelstham-Andronov school have been collected in the book *Theory of the Bifurcations of Dynamic Systems in the Plane*, by Andronov, Leontovitch, Gordon, and Mayer (Moscow, 1967). This book gives an exposition of the notion of *degree of structural instability* introduced in 1939 by Andronov and Pontriagin.

From the point of view of practice, the theoretical notion of dangerous and not dangerous stability boundaries in the parameter space, introduced by N.N. Bautin in 1949, is of outstanding importance. This notion is another refinement of the concept of structural instability. Passing through a dangerous stability boundary gives a clear-cut, and sudden, change in the qualitative behavior of the corresponding dynamic system, possibly irreversible. When the boundary of the parameter is not dangerous, the change in the qualitative behavior is progressive, continuous, and reversible. In engineering, it is fundamental to choose the parameters of a system sufficiently far from a dangerous boundary of stability in order to avoid premature failures or damage resulting from parameter variations due to aging of the systems and the random perturbations acting on it.

The notion of structural stability has an extension to dynamic systems described by:

(1) $dx/dt = f(x, y)$, $\mu dy/dt = g(x, y)$, $\mu > 0$

where x, y, are vectors, μ is a small vector parameter representing the parasitic elements of the system, and $f(x, y)$, $g(x, y)$ are bounded and continuous in the domain of interest of the phase space.

If $\mu = 0$, (1) reduces to a system of lower dimension:

(2) $dx/dt = f(x, y)$, $g(x, y) = 0$

For theoretical, as well as practical purposes, a fundamental problem consists in determining when the small terms $\mu \, dy/dt$, representing the effects of the parasitic elements, are negligible: In other words, when is the motion described by (1) sufficiently close to the motion described by (2), so that it can be represented by the solution of (2) defined for a lower dimension? Such a problem had its first general mathematical solution by A.N. Tikhonov (*Mat. Sbornik*, 31, 73 (1952), p. 575), but its most elegant solution was by Pontriagin (*Izv. Acad. Nauk Ser. mat.* 21 (1957) p. 605).

It is interesting to note that the formulation of this important problem has its origin in a discussion (1929) between Andronov and Mandelstham, related to the one time-constant electronic multivibrator. This oscillator is described by an equation such as (1), where x is a scalar (*dim x* = 1), the first equality being related to the multivibrator time constant, the second equality (*dim y* > 1) representing the effects of the parasitic inductances and capacitances. Without considering the parasitic elements such as parasitic capacitances and inductances, the multivibrator is nominally described by a one-dimensional, autonomous ODE, such as (2), where x is now a scalar (voltage). If it is required that $x(t)$ be a continuous function of time, then it was shown by Andronov that (2) does not admit any nonconstant periodic solution. Such a mathematical result is contrary to physical evidence, because the one time-constant multivibrator is known to oscillate with a continuous periodical waveform. In the Mandelstham-Andronov discussion of this paradox, the following alternative was formulated: either (a) the nominal model (2) is not appropriate to describe the practical multivibrator, or (b) it is not being interpreted in a physically significant way.

In 1930, Andronov showed that the two components of the alternative may be used to resolve the paradox, provided the space Y of admissible solutions (x, y) is properly defined. In fact, requiring that (x, y) be continuous and continuously differentiable leads to the conclusion that (2) is inappropriate on physical grounds, because a real multivibrator possesses several small capacitances and inductances. On the other hand, the dependance on parasitic elements can be alleviated by means of the second component (b) of the alternative, *i.e.*, with a generalization of the set of admissible solutions (x, y). In fact, if this set is defined as consisting of piecewise continuous and piecewise differentiable functions of (x, y), the first-order differential system (2) is supplemented by some appropriate jump conditions (Mandelstham conditions) for joining the various pieces of (x, y), and then (2) admits a piecewise continuous periodic solution. For the multivibrator example, the jump conditions reduce to the statement that the voltage of the multivibrator condenser should be a continuous function of time. Physically, this means that the electrical current charging this condenser should always be finite.

The Mandelstham-Andronov-Tikhonov-Pontriagin results regarding the problem (1) – (2) are of outstanding importance for two related problems:

- the *dimension reduction of a mathematical model* of a dynamic system;

- the theory of *relaxation oscillations.*

Central ideas of these results are now used in catastrophe theory which has rediscovered them.

The most important results of the Mandelstham-Andronov school, obtained until 1960, are given in the best book written to date about oscillations: A.A. Andronov, A.A. Witt, S.E. Khiakin, *Theory of oscillations* (Ed. Fiz. Mat. Literature, Moscow, 1937/ 1959). This book contains a high-level theoretical part, and a lot of applications to engineering and mechanics.

A6.4.3 More recent contributions to qualitative methods

One of the most important contributions is that of Yu.I. Neimark, who devoted his work primarily to map theory and its applications. His book *The Method of Mappings in the Theory of Nonlinear Oscillations* (Nauka, 1972, Moscow) presents this contribution. Another interesting contribution is that of N.N. Leonov on one-dimensional mappings which are piecewise continuous.

Concerning mappings, Neimark developed a method parallel to that adopted by Andronov for autonomous differential equations. Noteworthy are his studies of the critical cases in the Liapunov sense, various bifurcations under the influence of parameter variations when crossing through these critical cases, and the invariant manifolds of differential equations, as well as those of mappings. In particular, Neimark was the first (1968) to describe the bifurcation giving rise to an invariant closed curve from a complex focus of multiplicity one (improperly attributed to Hopf); the more general case of any multiplicity was studied in [Mira, 1969a] and [Gumowski and Mira, 1974].[1] This bifurcation corresponds to the crossing through a Liapunov critical case with two eigenvalues exp $(\pm i\varphi)$, $i = \sqrt{-1}$.[2]

Neimark has applied his results about mappings to many problems concerning ordinary differential equations, such as influence of parameters on periodic solutions, solutions of the piecewise linear equations, the method of the Poincaré's small parameter considered from the point of view of mappings, some extensions to the equations with discontinuous second member, the method of averaging, and the method of auxiliary mapping.

In the absence of necessary and sufficient conditions for the structural stability of m-dimensional dynamic systems, $m > 2$, the study of bifurcations was based initially on the conjecture that a system is structurally stable when the fixed (equilibrium) points and periodic solutions (orbits) are structurally stable and finite in number, when the set of nonwandering points consists of these stationary states only, and when all the stable and unstable manifolds intersect transversally. These sufficient conditions were formulated by Smale in 1960. Such systems are now known as Morse-Smale systems.

As a first step, Beliustina, Gavrilov, Mel'nikov, Neimark, and Shilnikov considered the bifurcations of these systems, in particular the birth of periodic solutions from a period orbit, the birth of a toroidal integral manifold from a periodic orbit, and the birth of a periodic solution from a trajectory (orbit) doubly asymptotic to a node-saddle ($m > 2$), or to a saddle-saddle ($m > 2$), or to a saddle ($m > 2$).

A second step was made by analyzing the bifurcations which transform a Morse-Smale system into a system having an enumerable set of periodic orbits. There are many such bifurcations.

1. These are references M14 and G30 in the book (Mira, 1987), which is in turn listed as M1 in the Bibliography of this book.

2. See (Babary and Mira, 1969), and (Mira, 1969b, 1970a, b, 1971), which are references B2, M15-18 in our M1.

Gavrilov, Afraimovitch, and Shilnikov have studied some of these bifurcations related to the presence of structurally unstable homoclinic, or heteroclinic curves associated to an equilibrium point, or to a periodic orbit for $m > 3$. The problem of structurally unstable homoclinic, or heteroclinic, structure was also considered by Neimark. The preceding results have contributed to the study of the popular *Lorentz-Saltzman differential equation* ($m = 3$) by Afraimovitch and Shilnikov. The bifurcations of this step correspond to certain transitions from order to chaos (or stochasticity).

For the framework of this second step, Shilnikov has introduced new notions such as *accessible and nonaccessible values of a bifurcation parameter.* He also specified the notions of *dangerous and nondangerous boundary of stability* in Bautin's sense for $m > 2$, and refined them with the complementary notions of *dynamically defined dangerous boundary and dynamically nondefined dangerous boundary* (cf. Shilnikov, *Dokl. Akad. Nauk, SSSR*, t. 224, no 5, 1975, pp. 1046 – 1049; Afraimovitch, Shilnikov, *Izv. Akad. Nauk, SSSR, Mat.*, t. 38, no 6, 1974, pp. 1248 – 1288).

A6.4.4   Contributions to analytical methods

For the Mandelstham-Andronov school, the analytical methods have been always an auxiliary tool for qualitative methods studies, or for the understanding of specific problems in physics or engineering. As early as 1932, Mandelstham and Papaleski used Poincaré's small parameter method in the study of nonlinear resonances, subharmonic resonances, synchronization phenomenon, and so on.

More recently, in 1956, I.G. Malkin published the book *Some Problems in the Theory of Nonlinear Oscillations* (Izd. Tekh. Teor. Lit. Moscow) about the method of small parameters. This book also gives a method of successive approximations in the nonanalytic case. Malkin has also published a study of critical cases in the Liapunov sense via analytical methods.

A6.4.5 Applications to physics and engineering

The theoretical results of the Mandelstham-Andronov school have practical applications that are very numerous and very important for purposes of both analysis and synthesis. They go beyond physics or engineering and concern all the sciences: population dynamics, economics, and so on.

Since the creation of the school, the applications have concerned all types of oscillators: electronic multivibrators, mechanical watches, or electro-mechanical machines, regulators using relays, steam engines, automatic flight with gyroscopes, radio-physics, quantum mechanics, dynamics systems with delay, and so forth. Some of these applications are described in the above-mentioned book of Andronov, Witt, and Khaikin.

Since 1952, when relations between the two groups (one in Gorki, the other in Moscow) of the Andronov school began to deteriorate, the group in Moscow has developed considerable activity in the automatic control field with contributions by Chetaev, Lur'e, Aizerman, Petrov, Meerov, Letov, Tsypkin, and others, and in optimization theory and practice with works of Pontriagin, Boltianski, Gamkrelidze, and Mischenko (for example, the maximum principle method).

A6.5 THE BOGOLIUBOV (OR KIEV) SCHOOL

This school has developed essentially the analytical methods, in the framework of the Institute of Mathematics of the Ukrainian Academy of Sciences. The foundation of their results is the classical method of perturbations, which has been generalized by this school to nonconservative systems. In 1932, the Krylov-Bogoliubov method gave a firm foundation to Van der Pol's studies of oscillators. The asymptotic method of Mitropolski improved the method further with asymptotically convergent series expansions. It is the same for the averaging method. Poincaré's method of small parameters is such that the full determination of the first harmonic and of the following harmonics of a periodical solution do not depend on the upper harmonics.

The contributions of this school are very great and concern systems with one, or several, degrees of freedom, the determination of periodical, or quasi- (and almost-) periodic solutions, transient regimes, integral manifolds, nonlinear resonances, subharmonic resonances, and synchronization phenomena. The work of this school is also oriented towards dynamic systems with pure delays. Apart from the considerations of analytical methods, from 1960 Sharkoskij has developed several fundamental studies on the singularities of one-dimensional noninvertible maps.

A6.6 POINCARÉ'S ANALYTICITY THEOREM

Consider the dynamic system

$$(3) \ dx/dt = f(x, t, \mu), \quad t > t_0, \quad x(t_0) = x_0$$

where all variables and parameters are real valued, t, t_0, and μ are scalars, x is a vector, and the functions f are at least continuous with respect to t and analytic with respect to x and μ, at least for $|x| < \bar{x}$ and $|\mu| < \bar{\mu}$.

Poincaré's well known analyticity theorem affirms that the solution of (3), with $\partial x_0 / \partial \mu = 0$, $|x_0| < \bar{x}$, can be expressed in the form of the series

$$(4) \ x = \sum_{i=0}^{\infty} x_i(t, x_0)\mu^i, \ x_i(0, x_0) = x_0, \text{ for } i = 0, i > 0$$

which is convergent for $|x_0| \leq \bar{x}_1 < \bar{x}, \ |\mu| < |\bar{\mu}_1| < \bar{\mu}$, and $0 < t - t_0 < \hat{t}(t_0, x_0, \mu)$. It is important to stress that $\hat{t}$ depends strongly on t_0, x_0 and μ, which is forgotten by many authors. It is worth noting that this theorem is frequently used incorrectly to prove the smooth dependence of the periodic solutions of (3) on the parameter μ, this dependence being presumed to follow from the analyticity of f with respect to x and $|\mu|$.

The limitation of $\hat{t}$ by the properties of f, other than smoothness, has a profound influence on any method of determining periodic solutions, because the validity of the Poincaré analyticity theorem in general cannot be extended to the asymptotic limit $t \to \infty$, or even to $\hat{t} \geq T$, where $T > 0$ is the period of the solution. The fundamental reason is that a periodic solution with a completely defined x_0 is parametrically imbedded in a general solution $x = F(t, t_0, x_0, \mu)$ with arbitrary x_0 such that $F = \bar{F}(x_0, \mu) = F_\infty(t, t_0)$ as $t \to \infty$. In general, it happens that $\bar{F}$ is a qualitatively different function of μ for different x_0, and for a fixed x_0 the precise dependence on μ is not known in advance. Moreover, the qualitative μ-dependence of F may exhibit a complex structure. At present no explicit conditions to be imposed on f are known which ensure that the μ-dependence of $\bar{F}$ is of a specific form for a fixed x_0. As a rule, different subharmonic resonances involve qualitatively different parametric dependences, and thus require qualitatively different types of convergent or asymptotic series expansions for their efficient quantitative description.

It is generally believed that the validity of the asymptotic method constitutes an extension of the Poincaré analyticity theorem, with usual μ-convergence replaced by asymptotic convergence. Many examples show that a mere asymptotic convergence is useless for the determination of a periodic solution unless, for a fixed difference between the exact and approximate solutions (say of 1%), the size of the approximation interval analogous to the convergence interval, $0 < t - t_0 < \hat{t}$, is not less than the period T. Unfortunately, if the differential equations are reduced to a quasi-linear standard form, which is the usual practice, and the expansion (4) does not apply with $\hat{t} < T$, then the asymptotic method either fails completely, or yields erroneous results.

The details of this important question are given in I. Gumowski's paper "Periodic steady states of dynamic systems and their smooth dependence on parameters" in *Theorie de l'iteration et ses applications*, Colloque C.N.R.S. no 332, Toulouse, 17 – 22 mai 1982, Editions du C.N.R.S., 1982, pp. 211 – 218.

A6.7 MYRBERG'S CONTRIBUTION

From 1958 to 1963, Myrberg published in the *Annales Academiae Scientiarum Fennicae* (ser. A, 256 (1958), 168 (1959), 336 (1963) a series of important papers concerning the bifurcation properties of the quadratic one-dimensional mapping *T*:

(5) $x_{n+1} = x_n^2 - \lambda$, $\quad n = 0, 1, 2,...;$ $\quad n = 0$, $\quad x = x_0$

where x is a real variable, and *y* is a real parameter. This map is non-invertible, the inverse having either a zero determination or two determinations. The singularities of the solution of (5) are constituted by:

- two fixed points satisfying $x = Tx$;

- the points of cycles of order (period) *k*, *k* = 2, 3, 4,..., such that $x = T^k x$, $x \neq T^l x$, $l < k$;

- their accumulation when *k* tends toward infinity; and

- more complex fractal sets resulting from complex accumulations of the preceding singularities.

Myrberg has shown with respect to (5) a series of important results for the theory of dynamic systems:

- All the bifurcation values of (5) occur in the interval

  $-1/4 < \lambda < 2$.

- The number N_k of all possible cycles having the same order *k*, and the number $N_\lambda (k)$ of bifurcation values giving rise to these cycles, increase very rapidly with *k*

  $k = 5$, $N_k = 6$, $N_\lambda (k) = 3$;
  $k = 10$, $N_k = 99$, $N_\lambda (k) = 51$;
  $k = 30$, $N_k = 35,790,267$, $N_\lambda (k) = 17,895,679$;
  $N_\lambda(k)$ and $N_k \to \infty$ if $k \to \infty$.

- The cycles with the same order differ from one another by the cyclic transfer of one of their points by *k* successive iterations by *T*. These cyclic transfers were defined by Myrberg using a binary code constituted by a sequence of (*k*-2) signs +, - (binary rotation sequence).

- For $\lambda < \lambda_{1(s)} = 1.401155189$, the number of singularities is finite (T is *Morse-Smale*). For $\lambda > \lambda_{1(s)}$ the number of singularities is infinite.

- The following cascades of bifurcations: attr. cycle $k2^i \rightarrow$ rep. cycle $k2^i$ + attr. cycle $k2^i$ + 1 ["attr. (rep.) cycle m" indicating an attractive (repulsive) cycle of order m] occur as λ increases, $i = 1, 2, 3,...$; k is a fixed value, $k = 1, 3, 4,....$

- For $i \rightarrow \infty$, the bifurcation values, for a given value of k, have a limit point $\lambda_{k(s)}$, $\lambda_{1(s)} < \lambda_{k(s)} < 2$.

- It is possible to classify all the binary rotation sequences via an ordering law.

- A binary rotation sequence can be associated to the λ value resulting from the accumulation of bifurcations such that $i \rightarrow \infty$, or $k \rightarrow \infty$. This rotation sequence satisfies the ordering law.

All these wonderful results have been systematically passed over in silence in the contemporary papers dealing with this subject, which has been very fashionable since 1978. Most of these results are now generally attributed to papers more recent than those of Myrberg.

Myrberg's results are particularly important in the theory of dynamic systems for the following reasons:

- The one-dimensional mapping (5) can be imbedded into a two-dimensional mapping with a unique inverse (diffeomorphism) such that:

$$(6) \quad x_{n+1} = x_n^2 - \lambda + z_n h(x_n, z_n), \quad z_{n+1} = bg(x_n, z_n)$$

where h, g are continuous and differentiable bounded functions such that the inverse of (6) is unique, and b is a parameter. When $b = 0$, (6) reduces to (5).

- A diffeomorphism such as (6) can be obtained via Poincaré's method of sections applied to the ODE (1), where x is a two-dimensional vector, and y is a scalar.

CHAOS IN DISCRETE DYNAMICAL SYSTEMS

Then in (6), b has μ^a, $a > 0$, as its order of smallness. The diffeomorphism (6) can be also obtained via the same method applied to a two-dimensional, nonautonomous differential equation with periodic coefficient in t (independent variable = time).

So, by imbedding into a higher-dimensional dynamic system, of continuous or discrete type, the bifurcation properties of (5), and in particular Myrberg's results, can be transferred to explain a series of complex dynamic behaviors, such as some chaotic behaviors. Then (5) appears as the simplest and the purest form without a parasitic element of a discrete dynamic system having fundamentally complex behavior encountered in higher-dimensional, and analytically more complex, systems. Myrberg's results have been used in [Gumowski & Mira, 1975], [Mira, 1976], and [Mira, 1987][1] to identify the organization of the bifurcations of (5) and (6). This organization is of fractal type and has been called a *box-within-a-box* bifurcation structure. In the case of (6), this fractal structure is of foliated type in the (λ, b) parameter plane.

A6.8 CONCLUSION

In the 19th century, Joseph Fourier wrote:

The study of Nature is the most productive source of mathematical discoveries. By offering a specific objective, it provides the advantage of excluding vague problems and unwieldy calculations. It is also a means to formulate mathematical analysis, and to isolate the most important aspects to know and to conserve. These fundamental elements are those which appear in all natural effects.

It is also worthy of note that the majority of scientists (including mathematicians) were not led to their discoveries by a process of deduction from general postulates or general principles, but rather by a thorough examination of properly chosen particular cases. The generalizations have come later, because it is far easier to generalize an established result than to discover a new line of argument. Generalization without any interest for the discovery is the temptation of a lot of mathematicians working now on dynamic systems.

---

1. These are references G32, M22 in our M1.

The famous mathematician Halphen often complained that nonessential generalizations are overcrowding the publication media; and this was 80 years ago.

The important development of the theory of dynamic systems, during this century, is due to the study of the natural phenomena encountered in these systems, and the rejection of nonessential generalizations, that is, to the study of concrete dynamic systems. The results obtained in the field of abstract dynamic systems have been possible only because of the foundations laid by the results from the field of concrete dynamic systems.

DOMAINS
OF THE FIGURES

The computed, two-dimensional figures from Chapters 4 through 7 (there are 95 of them) are based on numerous data, including:

- the map family

- the fixed parameteers

- the valued of the bifurcation parameter

- the main features included (attractors, saddle points, insets and outsets, critical curves, basins, etc.)

- the domain, a rectangle in the x-y plane

- the segments of critical curves which are iterated to create the curve segments shown in the figure etc.

The first four items of this list are actually given for each figure, either in the text near the figure or in the figure captions, or are evident in the figure.In this appendix, we list the domains of all the figures in Chapters 4 through 7. The actual MAPLE scripts which made these figures may be found on the companion CD-ROM.

Chapter 4 domains

Figure	xmin	xmax	ymin	ymax
4-1	-1.5	1.5	-1.1	1.8
4-2	-1.5	1.5	-1.1	1.8
4-3	-1.5	1.5	-1.1	1.8
4-4	-1.5	1.5	-1.1	1.8
4-5	-2.5	2.5	-2.5	2.5
4-6	-1.0	1.2	-0.8	1.5
4-7	-1.0	1.2	-0.8	1.5
4-8	-0.5	0.0	-0.55	-0.45
4-9	-1.2	1.2	-0.8	-1.5
4-10	-1.2	1.2	-0.8	-1.5
4-11	-1.2	1.2	-0.8	-1.5
4-12	-0.65	0.2	-0.61	0.2
4-13	-1.2	1.2	-0.8	-1.5
4-14	-1.2	1.2	-0.8	-1.5
4-15	-2.5	2.2	-2.0	2.5
4-16	-0.8	0.0	-1.05	-0.9
4-17	-2.0	2.0	-2.7	2.2
4-18	-0.8	0.4	-1.05	0.0

Chapter 5 domains

Figure	xmin	xmax	ymin	ymax
5-1	-1.5	1.5	-2.0	1.8
5-2	-1.5	1.5	-2.0	1.8
5-3	-1.5	1.5	-2.0	1.8
5-4	-0.3	1.1	-0.6	0.5
5-5	-0.3	1.1	-0.6	0.5
5-6	-1.5	1.5	-2.0	1.8
5-7	-1.5	1.5	-2.0	1.8
5-8	-1.5	1.5	-2.0	1.8
5-9	0.4	1.1	-0.6	0.5
5-10	0.4	1.1	-0.6	0.5
5-11	-1.5	1.5	-2.0	1.8
5-12	0.3	1.1	-0.6	0.5
5-13	0.3	1.1	-0.6	0.5

Chapter 5 domains (*continued*)

Figure	xmin	xmax	ymin	ymax
5-14	-1.5	1.5	-2.0	1.8
5-15	0.4	1.1	-0.6	0.5
5-16	-0.3	0.0	-0.6	-0.58
5-17	0.4	1.1	-0.6	0.5
5-18	-1.5	1.5	-2.0	1.8
5-19	0.4	1.1	-0.6	0.5
5-20	-1.5	1.5	-2.0	1.8
5-21	-0.3	0.0	-0.6	-0.58
5-22	-1.5	1.5	-2.0	1.8
5-23	-1.5	1.5	-2.0	1.8
5-24	0.4	1.1	-0.65	0.5
5-25	-1.5	1.5	-2.0	1.8
5-26	0.4	1.1	-0.6	0.51
5-27	0.4	1.1	-0.6	0.51
5-28	0.0	1.0	-0.6	1.0
5-29	-1.5	1.5	-2.0	1.8
5-30	0.4	1.1	-0.6	0.5
5-31	0.4	1.1	-0.6	0.5
5-32	-1.5	1.5	-2.0	1.8
5-33	-0.9	-0.45	-0.9	-0.45
5-34	-0.4	0.0	-0.65	-0.4
5-35	0.9	1.1	-0.65	-0.4
5-36	-0.2	0.2	0.9	1.1

Chapter 6 domains

Figure	xmin	xmax	ymin	ymax
6-17	-3.5	3.5	-4.0	8.0
6-18	-3.5	3.5	-4.0	8.0
6-19	-3.5	3.5	-4.0	8.0
6-20	-3.5	3.5	-4.0	8.0
6-21	-3.5	3.5	-4.0	8.0
6-22	-3.5	3.5	-4.0	8.0
6-23	-3.5	3.5	-4.0	8.0
6-24	-0.5	0.5	-0.5	1.2

Chapter 6 domains (*continued*)

Figure	xmin	xmax	ymin	ymax
6-25	0.1	0.3	0.5	0.9
6-26	0.181	0.275	0.69	0.83
6-27	0.255	0.261	0.734	0.740
6-28	-3.5	3.5	-4.5	8.0
6-29	-3.5	3.5	-4.5	8.0
6-30	-3.0	3.0	-4.0	4.0
6-31	-2.4	-1.0	-2.2	-1.9
6-32	-0.3	1.0	-2.15	-1.5
6-33	-2.25	0.75	-2.25	0.75
6-34	-3.3	3.3	-4.5	8.0
6-35	-2.4	-1.0	-2.2	-1.9
6-36	-0.15	0.65	-2.0	-1.0
6-37	-0.7	-0.5	-0.7	-0.2
6-38	-2.2	0.8	-2.2	0.5

Chapter 7 domains

Figure	xmin	xmax	ymin	ymax
7-16	0.422	1.1	0.05	0.98
7-17	0.70	0.72	0.83	0.90
7-18	0.713	0.721	0.871	0.885
7-19	0.7178	0.7183	0.8785	0.8800
7-20	0.7178	0.7183	0.8785	0.8800
7-21	0.422	1.1	0.05	0.98
7-22	0.14	0.89	0.65	1.0
7-23	0.48	0.74	0.82	0.9
7-24	0.54	0.58	0.822	0.828
7-25	0.54	0.58	0.822	0.828
7-26	0.422	1.1	0.05	0.98
7-27	0.422	1.1	0.05	0.98
7-28	0.50	1.0	0.45	1.0
7-29	0.0	1.0	0.05	1.0
7-30	0.0	1.0	0.0	1.0
7-31	0.0	1.0	0.0	1.0
7-32	0.5	1.0	0.0	1.0
7-33	0.0	1.0	0.0	1.0

BIBLIOGRAPHY

FREQUENTLY USED REFERENCES BY CODE

Books

A: V. I. Arnold, S.M. Gusein-Zade, and A.N. Varchenko, *Singularities of differentiable maps,* 2 vols., Boston: Birkhauser, 1985-1988.

B: A. F. Beardon, *Iteration of rational functions*, New York : Springer-Verlag, 1991.

GM1: I. Gumowski and C. Mira, *Dynamique Chaotique. Transition ordre-desordre*; Toulouse: Cepadues, 1980.

GM2: I. Gumowski and C. Mira, *Recurrences and discrete dynamic systems*, Lecture Notes in Mathematics 809, Berlin: Springer-Verlag, 1980.

M1: Christian Mira, *Chaotic Dynamics: From the One-dimensional Endomorphism to the Two-dimensional Diffeomorphism*, Singapore; Teaneck, NJ: World Scientific, 1987.

MGBC: C. Mira, L. Gardini, A. Barugola, and J.-C. Cathala, *Chaotic Dynamics, Two-dimensional Noninvertible Maps*, Singapore; Teaneck, NJ: World Scientific, 1996.

Articles

BB: C. Mira, D. Fournier-Prunaret, L. Gardini, H. Kawakami, and J. C. Cathala, Basin bifurcations of two-dimensional noninvertible maps: fractalization of basins, *Int. J. Bifurcations and Chaos* 4, 1994, pp. 343 – 381.

FMG: Daniele Fournier-Prunaret, Christian Mira, Laura Gardini, Some contact bifurcations in two-dimensional examples, *Proc. ECIT94, Opava September 1994*, in preparation.

G1: Laura Gardini, Homoclinic bifurcations in *n*-dimensional endomorphisms due to expanding periodic points, *Nonlinear Analysis TM& A* 23, pp. 1039 – 1089, 1994.

GARF: L. Gardini, R. Abraham, R. Record, D. Fournier-Prunaret, A double logistic map, *Int. J. Bifurcations and Chaos*, 4 (1994), 145 – 176.

HW. Hassler Whitney, On singularities of mappings of euclidean spaces. I. Mappings of the plane into the plane, *Ann. of Math.* (2) 62 (1955), 374 – 410.

KK. Kawakami, H., and K. Kobayashi, Computer experiments on chaotic solutions of $x(t+2) - ax(t+1) - x^2(t) = b$, *Bull. Fac. of Engin.*, Tokushima University, 16, 1979, pp. 29-46.

BIBLIOGRAPHY BY AUTHOR

Books

Abraham, Ralph H., Jerrold E. Marsden, and Tudor Ratiu, *Manifolds, Tensor Analysis, and Applications*, 2nd. ed., New York: Springer-Verlag, 1988.

Abraham, Ralph H., and Christopher D. Shaw, *Dynamics the Geometry of Behavior*, 2nd ed., Reading, MA: Addison-Wesley, 1992.

Bai-lin, Hao, *Elementary Symbolic Dynamics and Chaos in Dissipative Systems*, Singapore; Teaneck, N.J.: World Scientific, 1989.

Gumowski, I. and C. Mira, *Dynamique Chaotique. Transition ordre-desordre*; Toulouse: Ed. Cepadues, 1980.

Gumowski, I. and C. Mira, *Recurrences and discrete dynamical systems*, Lecture Notes in mathematics, Berlin: Springer-Verlag, 1980.

McCracken, M. and J. Marsden, *The Hopf Bifurcation*, Lecture Notes in Mathematics, Berlin: Springer-Verlag, 1976.

Mira, Christian, *Chaotic Dynamics: From the One-dimensional Endomorphism to the Two-dimensional Diffeomorphism*, Singapore ; Teaneck, NJ: World Scientific, 1987.

Mira. Christian, *Systemes asservis non lineares*, Paris: Ed. Hermes, 1990.

Mira, C., L. Gardini, A. Barugola, and J.-C. Cathala, *Chaotic Dynamics, Two-dimensional Noninvertible Maps*, Singapore ; Teaneck, NJ: World Scientific, 1996.

Articles

Allen, T., G. Oster, and D. Auslander, 1976.

Aronson, D. G., M. A. Chory, G. R. Hall, and R. P. McGehee, Chapter 2, Resonance phenomena for two parameter families of maps of the plane: uniqueness and nonuniqueness of rotation numbers, in: *Nonlinear Dynamics and Turbulence*, ed. E. Barenblatt, G. Iooss, and D. D. Joseph, 1983, pp. 35 – 47.

Aronson, D. G., M. A. Chory, G. R. Hall, and R. P. McGehee, Bifurcations from an invariant circle for two-parameter families of maps of the plane: a computer-assisted study, *Commun. Math. Phys.* 83, 1982, pp. 303 – 354.

Barugola, A., Determination de la frontiere d'une zone absorbante relative a une recurrence du deuxieme ordre a inverse non unique, *C.R. Acad. Sc. Paris*, Serie B, 290, 1980, pp. 257 – 260.

Barugola, A., Quelques proprietes des lignes critiques d'une recurrence du second ordre a inverse non unique. Determination d'une zone absorbante, *R.A.I.R.O., Analyse Numerique*, vol. 18, 1984, pp. 137 – 151.

Barugola, A., Sur certaines zones absorbantes et zones chaotiques d'un endomorphisme bidimensionnel, *Int. J. Nonlinear Mech.*, vol. 21, 1986, pp. 165 – 168.

Barugola, Alexandre, Sur certaines zones chaotique, *International Journal of Non-Linear Mechanics* 23, 1988, pp. 355 – 359.

Barugola, A. and J. C. Cathala, Sur les zones absorbantes et les zones chaotiques d'un endomorphisme bidimensionnel, *Ann. Telecommun.* 42, 1987, pp. 255 – 262.

Barugola A. and J. C. Cathala, Endomorphismes de dimension deux et attracteurs etranges, Actes du seminaire A.T.P. de Grenoble, in: M. Cosnard ed., *Traitement Numerique des Attracteurs Etranges*, Paris: Editions du C.N.R.S., 1985, pp. 81 – 106.

Barugola, A. and J. C. Cathala, An extension of the notion of chaotic area in two-dimensional endomorphisms, *ECIT 1992 Proceedings*. Singapore; Teaneck, NJ: World Scientific, to appear.

Barugola, A., J.C. Cathala, and C. Mira, Annular chaotic areas, *Nonlinear Analysis TM&A*, 10, 1986, pp. 1223 – 1236.

Bernussou, J., Hsu Liu, and C. Mira, Quelques exemples des solutions stochastiques bornees dans les recurrences du 2e ordre, *in: Transformations ponctuelles et applications*, Colloques Intern. du C.N.R.S., Paris, 1973, pp. 194 – 220.

Carcasses, Jean-Pierre, Determination of different configurations of fold and flip bifurcation curves of a one or two-dimensional map, *Int. J. Bifurcations and Chaos* 3, 1993, pp. 869 – 902.

Carcasses J.P., C. Mira, M. Bosch, C. Simo, J.C. Tatjer, Crossroad area-spring area transition. (I) Parameter plane representation, *Int. J. of Bifurcation & Chaos*, 1, 1991, pp. 183 – 196.

Cathala J.C., Determinations des zones absorbantes et chaotiques pour un endomorphisme d'ordre deux, Actes du Colloque, n.332 sur la theorie de l'iteration et ses Applications,Toulouse, 1982, Editions du C.N.R.S. pp. 91 – 98.

Cathala, J.C., Sur la dynamique complexe et la determination d'une zone absorbante pour un systeme a donnees echantillonnees decrit par une recurrence du second ordre, *R.A.I.R.O. Automatique*, 16, 1982, pp. 175 – 193.

Cathala, J.C., Absorptive area and chaotic area in two-dimensional endomorphisms, *Nonlinear Analysis T., M., & A.*, 7, 1983, pp. 147 – 160.

Cathala J.C., On the bifurcation between a chaotic area of T^k and a chaotic area of T, Proceedings ECIT84, R. Liedl and G. targonski eds, Lecture Notes in Mathematics 1163, Springer-Verlag , 1985, pp. 14 – 22.

Cathala, J.C., Bifurcations occurring in absorptive and chaotic areas, *Int. J. Systems Sci.*, 18, pp. 339 – 349, 1987.

Cathala J.C., Case where the influence domain of a stable attractor of a R^2-endomorphism is a multi-connected domain, *Proceedings ECIT87*, C. Alsine, J. Llibre, C. Mira, C. Simo, G. Targonski, R. Thibault eds, World Scientific 1989, pp. 161 – 166.

Cathala J.C., On some properties of absorptive areas in second order endomorphisms, *Proceedings ECIT89*, World Scientific, 19xx, pp. 42 – 54.

Cathala, J. C., H. Kawakami, and C. Mira, Singular points with two multipliers S1 = − S2 = 1, *Int. J. Bifurcations and Chaos*, 2, 1992, pp. 1001 − 1004.

Fournier-Prunaret D, Diffeomorphisme cubique de dimension deux. Frontiere des domaines d'attraction de deux attracteurs, *Comptes Rendu Acad. Sci.*, Paris, serie I, n.7, 1982, pp.727 − 730.

Fournier-Prunaret D., The bifurcation structure of a family of degree one circle endomorphisms. *Int. J. Bifurcation & Chaos*, 1, 1991, pp. 823 − 838.

Fournier-Prunaret D., C. Mira, Harmonic synchronization area of a nonautonomous system with quadratic non linearities, *Proceedings of the Int. Conference on Singular Behavior and Nonlinear Dynamics, Samos, 1988*, World Scientific, pp. 201 − 206.

Fournier-Prunaret, D., C. Mira, and L. Gardini, Some contact bifurcations in two-dimensional examples, *ECIT 1994 Proceedings*, to appear.

Fournier-Prunaret D., C. Mira, H. Kawakami, La chaine de liaison: image symbolique d'un ensemble comlet de communications entre feuillets du plan de biforcations d'un diffeomorphisme bidimensionnel, *Comptes Rendu Acad. Sci., Paris*, serie I, n.7, 1986, pp.315 − 320.

Gaertner, Wulf, and Jochen Jungeilges, A model of interdependent consumer behavior: nonlinearities in R^2, preprint, 1991.

Gardini, L., Some global bifurcations of two-dimensional endomorphisms by use of critical lines, *Nonlinear Analysis T, M,&A*, 18, 1991, pp. 361 − 399.

Gardini L., Global analysis and bifurcations in two-dimensional endomorphisms by use of critical curves, *Proceeding of ECIT91*, World Scientific, 1992a, pp. 114 − 125.

Gardini, Laura, Absorbing areas and their bifurcations, Report n 16, Univ. of Urbino, 1992b.

Gardini, L., On a model of financial crisis. Critical lines as new tools of global analysis, in: *Nonlinear Analysis in Economics and Social Sciences*, F. Gori, M. Galeotti, and L. Geronazzo, eds., Berlin: Springer-Verlag, 1993.

Gardini L., Homoclinic bifurcations in n-dimensional endomorphisms due to expanding periodic points, *Nonlinear Analysis T.M.&A.* 1994, 23, pp. 1039 – 1089.

Gardini, Laura and Ralph H. Abraham, About a map of coupled oscillators, Report n. 20, Univ. of Urbino, 1992.

Gardini, Laura, Ralph Abraham, Ronald J. Record, and Daniele Fournier-Prunaret, A double logistic map, *Intern. J. Bifurcations and Chaos* 4, 1994, pp. 145 – 176.

Gardini L., J.C. Cathala, C. Mira, Contact bifurcations of absorbing and chaotic areas in two-dimensional endomorphisms, in: *Proceedings ECIT92*, World Scientific, in press.

Gardini L., R. Lupini, One dimensional chaos in impulsed linear oscillators, *Int. J. Bifurcation & Chaos*, 1993, 3, pp. 921 – 941.

Gardini L., C. Mira, D. Fournier-Prunaret, Properties of invariant areas in two-dimensional endomorphisms, *Proceedings ECIT92*, World Scientific, in press.

Giraud, A., *Application des recurrences a l'etude de certains systemes de commande*, These de Docteur-ingenieur, Univ. of Toulouse, 1969.

Grebogi C., and E. Ott, Crises, sudden changes in chaotic attractors and transient chaos, *Physica* 7D, 1983, pp. 181 – 200.

Grebogi C., E. Ott, and J.A. Yorke, Fractal basin boundaries, long-lived chaotic transients, and unstable-unstable pair bifurcations, *Physical Review Letter* 50, 1983, pp. 935 – 938.

Guckenheimer, J., and G. Oster, Bifurcation behavior of population models, in: *The Hopf Bifurcation*, Marsden & McCracken, eds., Berlin: Springer-Verlag, 1976.

Guckenheimer, J., G. Oster, and A. Ipaktchi, The dynamics of density dependent population models, *J. Math. Biol.*, 4, 1977, pp. 101 – 147.

Gumowski, I. and C. Mira, Sur un algorithme de determination du domaine de stabilite d'un point double d'une recurrence non lineaire du deuxieme ordre a variables reeles, *C.R. Acad. Sc.*, (2) 260, 1965, pp. 6524 – 6527.

Gumowski, I. and C. Mira, Determination graphique de la frontiere de stabilite d'un point d'equilibre d'une recurrence non lineaire du deuxieme ordre a variables reelles. Application au cas ou les seconds membres de la recurrence ne sont pas analytiques, *Electronic Letters* 2, 1966, pp. 133 – 135.

Gumowski, I. and C. Mira, Sensitivity problems related to certain bifurcations in nonlinear recurrence relation, *Automatica*, 5, 1969, pp. 303 – 317.

Gumowski, I. and C. Mira, Cas critique, 1971.

Gumowski, I. and C. Mira, Sur la distribution, 1972.

Gumowski, I. and C. Mira, Determination des courbes, 1973.

Gumowski, I. and C. Mira, Solutions chaotiques bornees d'une recurrence ou tranformation ponctuelle du 2e ordre a inverse non unique, *C.R. Acad. Sc. Paris*, A285, 1977, pp. 477 – 480.

Gumowski, I. and C. Mira, Bifurcation destabilisant une solution chaotique d'un endomorphisme du deuxieme ordre, *C.R Acad. Sc. Paris*, A286, 1978, pp. 427 – 431.

Henon, M., A two dimensional mapping with a strange attractor, *Comm. Math. Physics*, 50, 1976, p. 69.

Kawakami, H., and K. Kobayashi, Computer experiments on chaotic solutions of $x(t+2) - ax(t+1) - x^2(t) = b$, *Bull. Fac. of Engin.*, Tokushima University, 16, 1979, pp. 29 – 46.

Li, T., and J. A. Yorke, Period three implies chaos, *Amer. Math. Monthly*, 82, 1975, pp. 985 – 992.

Lopez-Ruiz, R., and D. Perez-Garcia, Dynamics of two logistic maps with a multiplicative coupling, *Int. J. Bifurcations and Chaos* 2, 1992, pp. 421 – 425.

Lorenz, E., Computational chaos: a prelude to computational instability, *Physica D*, 35, 1989, pp 299 – 317.

Lorenz, E., Deterministic non-periodic flows, *J. Atmos. Sci.*, 20, 1963, pp. 130 – 141.

Marotto, J. R., Snap-back repellers imply chaos in R^n, *J. Math. Analysis Applic.* 63, 1978, pp. 199 – 223.

May, R. M., Biological populations with nonoverlapping generations: stable points, stable cycles, and chaos, *Science*, 186, 1974, pp. 645 – 647.

McDonald S.W., C. Grebogi , E. Ott, and J.A. Yorke, Fractal basin boundaries, *Physica* 17D, 1985, pp. 125 – 153.

Mira, C., Determination pratiques du domain de stabilite d'un point d'equilibre d'une recurrence non lineaire du deuxieme order a variable reelles, *C. R. Acad. Sciences Paris*, A261, 1965, pp 5314 – 5317.

Mira, C., Complex dynamics in two-dimensional endomorphisms, *Nonlinear Analysis T. M. and A.*, 4, 1980, pp. 1167 – 1187.

Mira C., and J.P. Carcasses, On the Crossroad area-saddle area and spring area transition, *Int. J. of Bifurcation & Chaos*, 1, 1991, pp. 641 – 655.

Mira C., J.P. Carcasses, M. Bosch, C. Simo, J.C. Tatjer, Crossroad area-spring area transition. (II) Foliated representation, *Int. J. of Bifurcation & Chaos*, 1, 1991, pp. 339 – 348.

Mira C., J.P. Carcasses, G. Millerioux, and L. Gardini, Plane foliation of two-dimensional noninvertible maps, *Int. J. Bifurcation and Chaos*, to appear.

Mira, C., D. Fournier-Prunaret, L. Gardini, H. Kawakami, and J. C. Cathala, Basin bifurcations of two-dimensional noninvertible maps: fractalization of basins, *Int. J. Bifurcations and Chaos* 4, 1994, pp. 343 – 381.

Mira, C. and I. Gumowski, Sur un algorithme de determination du domain de stabilite d'un point double d'une recurrence nonlineaire du deuxieme order a variables reelles, *C. R. Acad. Sciences Paris*, Serie A, (2) 260, 1965, pp 6524 – 6527.

Mira, C., and I. Gumowski, Sensitivity Problems related to certain bifurcations in nonlinear recurrence relation., *Automatica*, 5, 1969, pp 303 – 317.

Mira C., Y. Maistrenko, I. Sushko, On some properties of a piecewise linear two-dimensional endomorphism, *Int. J. Bifurcation and Chaos*, to appear.

Mira, C. and T. Narayaninsamy, On the behaviors of two two-dimensional endomorphisms: role of the critical curves, *Int. J. Bifurcatiions and Chaos*, 3, 1993, pp. 187 – 194.

Mira, C. and C. Rauzy, Fractal aggregation of basin islands in two-dimensional quadratic noninvertible maps, *Int. J. Bifurcations and Chaos*, to appear.

Mira, C. and J. C. Roubellat, Cas ou le domaine de stabilite d'un ensemble limite attractif d'une recurrence du deuxieme ordre n'est pas simplement connexe, *C.R. Acad. Sc. Paris*, A268, 1969, pp. 1657 – 1660.

May, R., and G. Oster, Bifurcations and dynamic complexity in simple ecological models, *Amer. Natur.* 110, 1975, pp. 573 – 599.

Roubellat, F., *Contributions a l'etude des solutions des recurrences non lineaires et applications aux systemes a donnees echantillonnees, These de Doctorat des Sciences*, 1969.

Takahashi, Y., and G. Oster, Models for age specific interactions in a periodic environment, *Ecological Monographs*, 44, 1974, pp. 483 – 501.

Ulam, S. M. and P. R. Stein, Non-linear transformation studies on electronic computers, *Rozprawy Metamatyczne*, 39, 1964, pp. 401 – 484.

Whitehead, R. R., and N. Macdonald, A chaotic mapping that displays its own homoclinic structure, *Physica* 13D, 1984, pp. 401 – 407.

Whitney, Hassler, On singularities of mappings of euclidean spaces. I. Mappings of the plane into the plane, *Ann. of Math.* (2) 62, 1955, pp. 374 – 410.

INDEX

A

absorbing area 41
absorbing interval 41
annular absorbing area 44
arborescent sequence 12
attractive fixed points 37
attractive focus 34
attractive node 35
attractors 7, 22, 37

B

basin 7, 22
bifurcation sequence 7
bifurcations 7, 23
boundary 7, 22, 169, 181
bounded set 181
box of the first kind 124

C

cascade 5, 171
catastrophic bifurcation 23
chaotic attractor 37
chaotic dynamic 226
cobweb construction 19
coincident inverses 6
complement 169
contact bifurcation 62
critical curve 4, 42
critical point 14, 30, 181
critical set 6
critical value 181
cusp point 103
cycle 16, 34
cyclic attractors 37
cyclic orbit 34

D

derivative 173
diagonal 16, 145
diffeomorphism 171
differential equation 186
domain 11
double logistic family 145
dynamic system 225
dynamical system 5
dynamics 225

F

fixed point 16, 33
flip bifurcation 23, 125
flows 5
fold 23
fold point 14
frontier 22, 142

G

generic 33

H

Hamiltonian 187
headland 92
heteroclinic 187
homoclinic 143, 187
homoclinic tangency 85

I

image 11, 29
immediate basin 92
initial point 7, 14
inset 45

interior 169
invariant 22
island 92
iterated map 6
iteration 5, 183

K
Koenig-Lemeray method 16

L
layer set 14, 29

M
map 11, 29, 171, 181
multiplicity 6, 14, 30
Myrberg map 22

N
noninvertible 5, 12, 29
nonlinear oscillations 225

O
orbit 14, 171
order 16

P
partial inverses 6, 14
period 34
period-doubling 125
periodic 16
periodic attractor 37
periodic orbit 34

Poincaré 186
portrait 7, 22
preimage 12, 29
prime period 34

R
range 11
repelling focus 36
repelling node 36
repellor 181

S
saddle 35
semi-cascade 5
separatrix 22
snap-back repellor 131
stable manifold 45
staircase method 19
stairway to chaos 5
state space 5
static attractor 37

T
tongue 62
topological closure 169
trajectory 7, 14, 171
transversal 42
trapping 22

Z
zone 6